LEARNING AND INSTRUCTION

A SERP RESEARCH AGENDA

Panel on Learning and Instruction
Strategic Education Research Partnership
M. Suzanne Donovan and James W. Pellegrino, editors

Division of Behavioral and Social Sciences and Education

NATIONAL RESEARCH COUNCIL
OF THE NATIONAL ACADEMIES

THE NATIONAL ACADEMIES PRESS
Washington, D.C.
www.nap.edu

THE NATIONAL ACADEMIES PRESS
500 Fifth Street, N.W. • Washington, DC 20001

NOTICE: The project that is the subject of this report was approved by the Governing Board of the National Research Council, whose members are drawn from the councils of the National Academy of Sciences, the National Academy of Engineering, and the Institute of Medicine. The members of the committee responsible for the report were chosen for their special competences and with regard for appropriate balance.

This study was supported by Grant No. R305U000002 between the National Academy of Sciences and the U.S. Department of Education; Grant No. 00-61980-HCD from the John D. and Catherine T. MacArthur Foundation; Grant Nos. 200200171 and 20030091 from the Spencer Foundation; and Grant No. B7070 from Carnegie Corporation of New York. Any opinions, findings, conclusions, or recommendations expressed in this publication are those of the author(s) and do not necessarily reflect the views of the organizations or agencies that provided support for the project.

Library of Congress Cataloging-in-Publication Data

Learning and instruction : a SERP research agenda : panel on learning and instruction / M. Suzanne Donovan, and James W. Pellegrino, editors.
    p. cm.
"Division of Behavioral and Social Sciences and Education."
  ISBN 0-309-09081-4 — ISBN 0-309-52762-7
  1. Education—Research—United States. 2. Reading—United States. 3. Mathematics—Study and teaching—United States. 4. Science—Study and teaching—United States. I. Donovan, Suzanne. II. Pellegrino, James W. III. National Research Council (U.S.). Panel on Learning and Instruction.
  LB1028.25.U6L43 2004
  370'.7'2—dc22
                       2003020850

Additional copies of this report are available from the National Academies Press, 500 Fifth Street, N.W., Lockbox 285, Washington, DC 20055; (800) 624-6242 or (202) 334-3313 (in the Washington metropolitan area); Internet, http://www.nap.edu.

Printed in the United States of America.
Suggested citation: National Research Council (2004). *Learning and Instruction: A SERP Research Agenda*. Panel on Learning and Instruction. M. Suzanne Donovan and James W. Pellegrino, editors. Division of Behavioral and Social Sciences and Education. Washington, DC: The National Academies Press.

THE NATIONAL ACADEMIES
Advisers to the Nation on Science, Engineering, and Medicine

The **National Academy of Sciences** is a private, nonprofit, self-perpetuating society of distinguished scholars engaged in scientific and engineering research, dedicated to the furtherance of science and technology and to their use for the general welfare. Upon the authority of the charter granted to it by the Congress in 1863, the Academy has a mandate that requires it to advise the federal government on scientific and technical matters. Dr. Bruce M. Alberts is president of the National Academy of Sciences.

The **National Academy of Engineering** was established in 1964, under the charter of the National Academy of Sciences, as a parallel organization of outstanding engineers. It is autonomous in its administration and in the selection of its members, sharing with the National Academy of Sciences the responsibility for advising the federal government. The National Academy of Engineering also sponsors engineering programs aimed at meeting national needs, encourages education and research, and recognizes the superior achievements of engineers. Dr. Wm. A. Wulf is president of the National Academy of Engineering.

The **Institute of Medicine** was established in 1970 by the National Academy of Sciences to secure the services of eminent members of appropriate professions in the examination of policy matters pertaining to the health of the public. The Institute acts under the responsibility given to the National Academy of Sciences by its congressional charter to be an adviser to the federal government and, upon its own initiative, to identify issues of medical care, research, and education. Dr. Harvey V. Fineberg is president of the Institute of Medicine.

The **National Research Council** was organized by the National Academy of Sciences in 1916 to associate the broad community of science and technology with the Academy's purposes of furthering knowledge and advising the federal government. Functioning in accordance with general policies determined by the Academy, the Council has become the principal operating agency of both the National Academy of Sciences and the National Academy of Engineering in providing services to the government, the public, and the scientific and engineering communities. The Council is administered jointly by both Academies and the Institute of Medicine. Dr. Bruce M. Alberts and Dr. Wm. A. Wulf are chair and vice chair, respectively, of the National Research Council.

www.national-academies.org

Contents

Preface

This report was prepared by the Panel on Learning and Instruction as a companion document to the report *Strategic Education Research Partnership* (SERP), prepared by the SERP Committee. The committee's report provides the vision for a new organization, program, and partnership that would allow education research and development (R&D) to be linked to, and embedded in, educational practice. This report puts flesh on the bones of the proposal for an R&D program focused on educational practice.

The panel and the committee worked simultaneously and separately. Because the staff to both groups was the same and the panel chair attended all committee meetings, the panel was kept informed of the committee's work throughout. The vision of the possible in this report assumes the research and development infrastructure that is proposed in the committee's report.

The task given to the panel was in some respects an uncomfortable one: to create a research and development agenda that would produce work that is genuinely useful to classroom teachers. To accomplish our mission, we needed to identify both the problems of practice that are particularly important to tackle early on, and cases of existing R&D that are particularly promising for improving classroom practice if carried further. This required that we make the case that some problems are more urgent than others, and some approaches to solutions show more promise than others.

This was a challenge for two reasons. First, a thorough review of all the literature that is relevant to K-12 education was beyond our scope and time frame. Instead, we relied on the syntheses of literatures done by previous National Research Council committees and committees formed elsewhere (including the RAND Corporation and the National Institute of Child

Health and Human Development), as well as on the breadth and depth of knowledge brought by panel members. While we believe our agenda is carefully considered and, if carried out, would genuinely improve educational practice, we are also aware that a different group of individuals might have produced a different agenda with equal merit.

Second, those who engage in research often are inspired by the interesting questions that still remain to be answered. Research findings are often viewed as an opening to new questions. But those in the world of educational practice require closure: When can we say that we know something with enough confidence to change the way we teach? Closure is anathema to researchers whose training leads them to question whether they have seen only a small piece of the puzzle, and whether the next piece will show them a picture different from what they had imagined.

If the researchers on the panel were somewhat hesitant, however, the teachers were less so. They work with children whose success in school and opportunities for the future they care deeply about. Their message was clear: we need access to the best research-based knowledge available, even when there are questions that remain unanswered, and even when more work remains to be done. If falling short of certainty means we do not identify and build on research and development with high potential for improving practice, then we will miss opportunities to improve student achievement that are sorely needed.

What we offer in this report, then, is an "illustrative" agenda: the best effort of a group of practitioners and researchers to identify research and development opportunities and directions that can support and sustain improvements in educational practice. We believe the overall framework we offer is one that can productively guide the development of a use-inspired R&D agenda on learning and instruction, even if any specific line of research we propose is replaced with another deemed to be of higher priority. We also believe that the lines of research outlined in this report have considerable potential for enhancing educational practice in the near term and for the more distant future. Most of all, our hope is that the agenda we have outlined, and its framing, will be helpful as a point of departure for a more extended and multifaceted discussion and decision-making process regarding SERP research priorities that will become a standard feature of education R&D in the future.

The panel is grateful to the many people who contributed to this report. The financial support of our sponsors at the Department of Education, the Spencer Foundation, the John D. and Catherine T. MacArthur Foundation, and Carnegie Corporation of New York was essential. Our thanks to C. Kent McGuire, former assistant secretary of education research and improvement, and to his successor and now director of the National Institute for Education Sciences, Grover J. Whitehurst; thanks likewise are due to Valerie Reyna, Mark Constas, and Sue Betka. We are grateful to Daniel Fallon, director of the education program at Carnegie Corporation, his predecessor Vivien Stewart, and colleague Karin Egan; Ellen Condliffe Lagemann, former president of the Spencer Foundation; and Paul Goren, vice president of the Spencer Foundation and before that education officer at the MacArthur Foundation.

The panel benefited greatly from the contributions of colleagues who shared their insights about the challenges of creating an agenda that could link research and practice more successfully than in the past. Catherine Snow of Harvard University and vice chair of the SERP committee talked with the panel during a formative stage, and her ideas and insights contributed significantly to the chapter on reading. Thomas Glennan of the RAND Corporation also spoke with the panel at the outset, sharing his extensive experience on issues of linking research with development and with practice. Kenneth Koedinger of Carnegie Mellon University and Kathleen Metz of the University of California, Berkeley, responded very helpfully to requests by the panel for further information on their work. Laura Cooper, assistant superintendent in Evanston Township High School in Evanston, Illinois, and a member of the SERP committee, provided feedback to the panel from her extensive experience in linking research with practice in school settings.

At the National Academies, Alexandra Wigdor provided suggestions and valuable feedback as the panel shaped its report, giving us the benefit of her vast experience throughout the process. Carole LaCampagne, a study director for mathematics projects, provided the panel with assistance on the mathematics chapter. Several project assistants supported the project. Shirley Thatcher supported the project at its inception, Allison Shoup carried the project through several meetings and a first draft of the report, and Neale Baxter and Elizabeth Townsend saw the report through its final stages. The panel is grateful to all for their hard work and good spirit. Final editing of the report was done by Christine McShane, with her trademark skill and careful attention to detail.

This report has been reviewed in draft form by individuals chosen for their diverse perspectives and technical expertise, in accordance with procedures approved by the National Research Council's Report Review Committee. The purpose of this independent review is to provide candid and critical comments that will assist the institution in making its published report as sound as possible and to ensure that the report meets institutional standards for objectivity, evidence, and responsiveness to the study charge. The review comments and draft manuscript remain confidential to protect the integrity of the deliberative process. We wish to thank the following individuals for their review of this report: Sherri Andrews, General Studies, North Carolina School of the Arts, Winston-Salem; Nicholas A. Branca, Mathematical and Computer Sciences, San Diego State University; James R. Brown, superintendent, Glendale Unified School District, CA; Williamson M. Evers, Hoover Institution, Stanford University; Richard M. Felder, Department of Chemical Engineering, North Carolina State University; Henry M. Levin, Teachers College, Columbia University; Marcia C. Linn, Graduate School of Education, University of California, Berkeley; Barbara Schneider, Sociology and Human Development, The University of Chicago; and Neil J. Smelser, Department of Sociology, University of California, Berkeley.

Although the reviewers listed above have provided many constructive comments and suggestions, they were not asked to endorse the conclusions or recommendations nor did they see the final draft of the report before its release. The review of this report was overseen by Richard J. Shavelson, School of Education, Stanford University, and William H. Danforth, chancellor emeritus and vice chairman, Board of Trustees, Washington University. Appointed by the National Research Council, they were responsible for making certain that an independent examination of this report was carried out in accordance with institutional procedures and that all review comments were carefully considered. Responsibility for the final content of this report rests entirely with the authoring committee and the institution.

Finally, we would like to sincerely thank all of the panel members who generously contributed their time and intellect to the completion of this project. Generating a comprehensive research agenda on learning and instruction in a short period of time represents an extraordinary challenge, requiring individuals with different backgrounds and interests to select among and prioritize a broad array of possible topics and issues, all of which may have purchase for improving educational practice. Throughout the process, the panel members displayed an extraordinary ability to make choices and maintain a sense of purpose and focus. Simultaneously, they showed a strong commitment to drawing on each other's expertise and the collective knowledge of the group. It has been both a professionally stimulating and personally gratifying experience to work with the members of this panel and everyone at the National Research Council associated with the SERP effort.

Jim Pellegrino, *Panel Chair*
Suzanne Donovan, *Study Director*

LEARNING AND
INSTRUCTION
A SERP RESEARCH AGENDA

Executive Summary

The Strategic Education Research Partnership (SERP) is a proposed large-scale, sustained program of education research and development (R&D). Its purpose is to provide a powerful knowledge base, derived from both research and the study of practice, that can support the efforts of teachers, school administrators, colleges of education, and policy officials to improve student learning. The defining feature of SERP is the tight coupling of research and practice: effective educational practice is both the goal and the subject of the R&D program. Program priorities will be negotiated among practitioners, policy makers, and researchers, and practitioners will work in collaboration with researchers in the execution of the program.

The SERP committee, in its report *A Strategic Education Research Partnership*, proposes an organizational design for carrying the SERP mission forward. At the heart of that design are networks through which focused, coordinated, and sustained programs of R&D are carried out, often in schools or school districts that serve as field sites. At full function, SERP would have multiple networks. In the start-up years, the committee proposes three, one of which would be a learning and instruction network.

To provide a more detailed vision of the program a SERP network might undertake, a separate SERP panel was convened to design an illustrative agenda for the learning and instruction network. To narrow program focus, the panel asked two questions:

1. *Are there examples of rigorous research and development efforts that already show impressive gains in student achievement in research trials?* These would provide

the starting point for R&D that is intended as one of the hallmarks of SERP: following promising program outcomes with research on the circumstances under which the results are obtained, the feasibility of the intervention in the classroom and the school context, the teacher knowledge and support required for success, and the organizational factors that influence outcomes. We refer to these as "downstream" cases because the work in these areas has already traveled some distance toward classroom usability.

2. *Are there pervasive problems of practice that are widely recognized as critical, but for which the knowledge base is too weak to guide instructional practice?* We refer to these as the "upstream" cases because the work is still in early stages. Since there are few promising interventions in these areas at present, an impact on practice is likely to require more time.

The panel recommends three areas for focus: reading, mathematics, and science. In both mathematics and reading, the proposed downstream work would address learning in the elementary years. In science, in contrast, the downstream work is in physics—a subject generally taught in high school—because of the strength of the research base. The upstream research proposed for the three domains focuses on reading comprehension, algebra, and the sequencing and content of science instruction across the school years.

· ·

READING

EARLY READING

There is an unusual degree of consensus regarding the goals of early reading instruction, as well as a fairly solid research base both on the contributors to success in achieving those goals and on assessments to predict reading difficulties. The targets of early reading instruction are therefore fairly clear.

Still, many children in U.S. schools are not learning to read well and, in many classrooms, teaching practices have not been influenced by research knowledge. In the panel's view there is a gap between the knowledge base on the contributors to success in early reading and the knowledge base on effective instruction and teacher knowledge requirements. The goal must be to move from principles of good literacy instruction to the practices and programs that will support it. We propose three initiatives for early reading:

1. development and testing of instructional approaches to narrowing the gap in early reading preparedness, with emphasis on programs to enrich oral language skills for children at ages 3, 4, and 5 and for native and nonnative English speakers;
2. the development and testing of models of integrated reading instruction for early elementary grades; and
3. research on the knowledge requirements for teachers of early reading, coupled with the development and evaluation of teacher education programs and tools and the development and validity testing of assessment measures to evaluate their effectiveness.

READING COMPREHENSION

Many students who learn to read successfully nonetheless do poorly at reading comprehension. There is remarkably little instruction to support reading comprehension in schools, perhaps because there is little science-based understanding of how comprehension builds or how to support its development over the years.

We propose four initiatives for reading comprehension:

1. research and development of formative and summative assessments of reading comprehension that capture the multiple components of effective comprehension and that span the school years;
2. research and development of instructional materials, protocols, and supports at different grade levels for teachers who are learning to use metacognitive strategy instruction in the classroom;
3. research on the instructional practices of teachers whose students "beat the odds" in their reading comprehension performance, with companion efforts to test the emerging hypotheses and incorporate tested

practices into systematic instructional programs that can be tested experimentally across a range of students;

4. research to support the development of benchmarks for reading comprehension across the school years.

· ·

MATHEMATICS

EARLY MATHEMATICS

Investment in recent decades by federal agencies and private foundations has produced a wealth of knowledge on the development of mathematical understanding and numerous curricula that incorporate that knowledge. Existing evaluations of some of these curricula suggest the potential to substantially improve student achievement outcomes, in some cases raising the performance of disadvantaged students to the level of their more advantaged counterparts. But adequate research has not been done to independently and rigorously evaluate the programs, to compare outcomes across these programs and with more traditional curricula, to study the teacher knowledge requirements the programs entail, and to consider the requirements for taking the programs to scale.

We propose three initiatives for early mathematics:

1. development of early mathematics assessments that capture the range of understanding and skills involved in mathematical proficiency, with companion research and development efforts aimed at identifying and providing the supports needed by teachers to use assessments effectively in teaching;

2. research and development on the knowledge required to teach elementary mathematics, on alternative approaches to teacher education that would support that knowledge development, and on the teacher supports required to take promising curricula to scale; and

3. independent evaluation and comparison of curricular approaches to the teaching of number and operations that vary on distinct and theoretically impor-

tant dimensions, with further research and development of component features of particular interest.

ALGEBRA

Algebra is crucial to the development of mathematical proficiency because it functions as the language system for ideas about quantity and space and is foundational for much other mathematics. However, there is currently little agreement regarding the content that should be included in the algebra curriculum or the instructional approach that is most effective.

We propose four initiatives aimed at improving algebra learning and instruction:

1. research and development on alternative approaches in the teaching and learning of algebra, with controlled experimentation at the level of particular program features, with a companion effort to extend existing curricula in promising directions;
2. research on the knowledge of mathematics needed to teach algebra effectively at different grade levels and research and development on effective teacher education interventions;
3. research and development of algebra assessments for the range of grade levels and the range of assessment purposes; and
4. study of students' algebra proficiency over time with the introduction of algebra as a K-12 topic.

• •

SCIENCE

PHYSICS

The existing knowledge base in physics education is relatively advanced. It includes the development and testing of exemplary instructional programs, with outcomes that suggest the possibility of a much deeper conceptual understanding of the subject, and by a broader range of students than are typically successful at physics today. We propose three initiatives designed to take research-based knowledge into the classroom:

1. refining the knowledge base on instructional programs to better distinguish the programs and their outcomes from each other and to identify the conditions and the contexts that typically accompany success;

2. research on the requirements to take promising physics curricula to scale in different school contexts, with companion efforts to develop supports for independent use of a curriculum;

3. research on teacher knowledge requirements for effective use of a curriculum and how that knowledge builds with teacher learning opportunities and experience.

SCIENCE EDUCATION ACROSS THE SCHOOL YEARS

International and national test scores highlight the weakness of K-12 science education in the United States. There are some indications that the absence of an agreed-on content for K-12 science instruction and the broad coverage of topics typical in science textbooks have led to weak development of scientific concepts over the school years. Improvement will require that choices be made that narrow the set of topics to be covered, and that instructional approaches to developing a deeper understanding of scientific concepts be identified or developed and evaluated for their outcomes. We propose three initiatives toward that end:

1. development and evaluation of integrated learning-instruction models aimed at identifying a productive organizing core for school science across the grades, with component curriculum and assessment research and development that extends existing promising efforts;

2. research and development on teacher knowledge requirements to effectively work with the curriculum under study; and

3. an ongoing effort to identify the feasibility and required commitments to achieve standards for science achievement.

ENSURING RESEARCH QUALITY AND IMPACT

The research agenda that we propose would represent a major, long-term investment in education research and development. We note in conclusion some of the features of the SERP organization, and of the agenda, that support both the quality of that investment and the likelihood that it will have an impact on practice.

To ensure quality, method is carefully matched to the question under study. No single methodology is favored in the agenda. We propose programs of research motivated by significant questions of educational practice and draw on the variety of methodologies required both to understand processes and to rigorously test outcomes.

The proposed program would have the ability to investigate empirically the problems of instructional practice in SERP field sites at which high-quality data can be collected longitudinally. The agenda calls for careful attention at the start to the development and testing of outcome measures, work that is often overlooked in research, but that is critical to the quality and interpretability of the results.

Since much of the research and development will take place in field sites, the R&D teams will be able to document critical elements of context; important differences among students from one classroom or school to the next can substantially affect outcomes. At the same time, no matter how carefully a single study is designed, confidence in results can be ensured only when they are replicated. Hence, the SERP agenda emphasizes both replication of research findings and testing of their range of applicability.

Impact requires actively seeking out high-quality research that is important for educational practice and building on it. But impact also requires the design of research studies that can take the knowledge from practice and incorporate it into testable propositions that can be shared publicly. Much of what teachers learn from repeated observation of student learning and response to instruction is never formally articulated, tested, or shared with others in their professional community. Proposed research designed to learn from practice can formalize the knowledge of teachers, subject it to testing, and make those results available to be publicly shared and scrutinized.

And finally, to ensure quality oversight, the SERP research would be subject to both internal and external scrutiny and critique.

The work we propose here has the potential to substantially improve the knowledge base to support teaching and learning by pursuing answers to questions at the core of teaching practice. It calls for a linking of research and development—of instructional programs, assessment tools, teacher education programs and materials—that is now rare. It would bring research to bear on the problems of educational practice. And just as importantly, it would bring the problems of practice to the agenda of research.

1 Introduction

The Strategic Education Research Partnership (SERP) proposes a large-scale research and development program that is tightly coupled with educational practice. The SERP mission is to develop a well-organized and powerful knowledge base, derived from both research and the study of practice, that supports the efforts of teachers, school administrators, colleges of education, and policy officials to improve student learning. The proposed initiative is fully described in *Strategic Education Research Partnership* (National Research Council, 2003).

At the heart of the SERP enterprise are networks through which focused, coordinated, and sustained programs of research and development are carried out collaboratively by practitioners and researchers. The proposed inaugural networks include one on learning and instruction and one on schools as organizations, with a network on education policy to follow. The broad priorities for the research and development networks are to be set by an advisory board, a group of distinguished practitioners, policy makers, and researchers who will together define the issues of greatest importance and with greatest potential payoff from focused R&D. Defining the specifics of the research and development program is the job of the SERP director and the researchers and practitioners who lead each network.

To illustrate the kind of work that the SERP enterprise will undertake, a panel of practitioners and researchers was convened to design an illustrative agenda for a prototypical network on learning and instruction. The panel members were chosen in part because they themselves have worked at the intersection of research and practice.

In the work presented here, we are to some extent simulating the role of the future SERP advisory board and network leadership. We focus broadly on target questions and on the panel's view of the nature of the work to be done. Ultimately the choice of topics and the finer detail of project definition will fall to those responsible for charting a course of action.

The goal of the panel was narrowly defined by the task of agenda development. The broad purposes and organizational structure of the SERP enterprise, and the practical challenges of creating a successful organization, were questions addressed by the SERP committee. The panel went about its work assuming the existence of a SERP organization like that proposed in the committee's report. The two reports, then, can be viewed as companion documents.

The decision to focus on learning and instruction was a function of the state of the research base from which we could draw. The National Research Council has in recent years produced syntheses of the research literature on human learning (National Research Council, 2000) and on assessment of learning (National Research Council, 2001b), as well as subject-specific syntheses in reading and in mathematics (National Research Council, 1998, 2001c). These and other explorations of the knowledge base on learning and instruction (National Institute of Child Health and Human Development, 2000; RAND 2002a, 2002b) provide a rich foundation on which our effort could build. But our focus is not intended to suggest preeminence of the learning and instruction network. Indeed, *it is in combination with the work of other networks—in particular, the proposed network on schools as learning organizations—that the R&D on learning and instruction will best be able to influence practice.*

The goal of the SERP initiative is to improve student learning. The panel reached a critical decision at the outset that gives structure to a learning and instruction research agenda that will further that goal: to *focus on practice*. How effectively students learn in school is in large part a function of the effectiveness of educational opportunities teachers provide to students, as well as the transactions between the teacher and the students that make those experiences productive. The problem before the panel, then, was to consider how research and development can support the teacher's effectiveness in providing—and helping students make use of—powerful learning opportunities. This means that the point of departure in defining the research and development program is not the intriguing questions that are at the frontiers of disciplines relevant to education (such as cognitive science and developmental psychology) that might have important implications for practice. Rather, it is the set of questions for which a classroom teacher needs answers in the conduct of instructional practice. Although the two would intersect in what has been referred to as Pasteur's quadrant (Stokes, 1997), the program of work will be shaped differently if practice is made central.

FOCUS ON PRACTICE

The potential role for research and development in supporting practice is perhaps seen more easily in health care, for which productive links between research and practice have been established over the past century. At the core of medical practice is the doctor's decision making, which is grounded in a knowledge base about the human body, about disease, and about interventions that can promote health and cure disease. That knowledge is embedded in the training of medical professionals; in the tools, protocols, and interventions that are standard in medical practice; and in the infrastructure for communicating to medical practitioners changing standards of practice. The generation of knowledge and its effective incorporation into training, tools, and protocols are not all that is required. But there can be no question that they are a necessary condition of sound practice.

A parallel core knowledge base for educational practice is that on learning, instruction, and subject matter. It is this knowledge base with which a learning and instruction network must concern itself. Broadly speaking, its task would entail research and development on the questions at the core of classroom practice: on the generation of knowledge of how students learn (both generically and in the contexts of particular content areas), the elaboration of that knowledge at the level of detail required for classroom practice, and the incorporation of that knowledge into tested curricula and assessments, educational tools, teaching protocols, and teacher education programs.

Medicine and education differ in a great many respects. Teaching, for example, focuses far more on promoting growth in a diverse group of students simultaneously, while medicine focuses on curing the individual case.[1] The analogy is illuminating, however, regarding the relationship between research and practice. Physicians regularly exercise a great deal of judgment and expertise in the conduct of a practice that brings new phenomena, or new presentations of familiar phenomena, daily. Yet we rely on a powerful link between research and practice to ensure that the knowledge stores from which the physician draws in the course of practice are well stocked and reasonably current. With the pace of change in medical research, ensuring currency is not a trivial problem (McGlynn et al., 2003; Avorn et al., 1982). But the need to grapple with that problem is considered central to maintaining professional standards.

In any field, research informs but does not define professional practice. Teaching, like many other professional practices, is a highly complex, multidimensional enterprise. It draws on standards of practice, professional education, background knowledge, tradition, and the personal characteristics and intuitions of the teacher. Research can play a major role in shaping these influences—for example by influencing professional education and standards of practice. But it is more likely to play that role if it is focused on the problems of classroom teaching and learning.

In medicine we would recognize as common professional practice the pursuit of answers to a core set of questions:

- How is the patient's health currently, and how does this compare to developmental expectations?
- What should the practitioner do to promote health and prevent disease or medical problems given current health and development?

When a medical problem exists:

- What are the patient's symptoms?
- What is the patient's personal history?

---

[1]Even this difference is a matter of degree and not kind, however. Recently there has been greater attention to "population health" issues that focus on characteristics of communities that produce or reduce medical risk (see National Research Council, 2001d), and the individual education plans required for special education placements focus on the learning needs and challenges of individual students.

- What tests can aid diagnosis?
- What diagnosis is warranted given the symptoms, history, and test results?
- What treatments are available?
- What is the best match between the patient, symptoms, and treatments?
- How is the patient responding to treatment?

While experience, intuition, and traditions of practice all contribute to a physician's approach to each question, research-based knowledge is at the core of professional practice. Indeed, the professionalization of medicine in the first half of the 20th century can be viewed as a transformation that placed research-based knowledge at the center of practice. Most certainly, knowledge flows in both directions; observations from medical practice generate many questions and hypotheses that fuel research. New discoveries quite often emerge as unanticipated by-products of medical treatments. But research is required to answer the questions and test the hypotheses and insights generated by practice.

Using the example of medicine, we can view a teacher's practice as organized around a set of core questions about human learning and instruction. These questions focus on the normal course of development and learning, as well as on diagnosing and responding to student problems in mastering new concepts and acquiring new knowledge and skills. And as in medicine, the questions provide a schema for approaching practice, as well as a set of dimensions on which knowledge must be supported through preparatory education for professional practice. The questions can be asked for any subject or topic that is taught:

- What should students know or be able to do?
- What common understandings and preconceptions do students bring to a topic?
- What is the expected progression of understanding, and what are the predictable points of difficulty or hurdles that must be overcome?
- What instructional interventions (curricula, instructional activities, etc.) can support the desired student learning?

- What general and discipline-specific norms and practices best comprise and support student learning?
- And finally, how can students' understanding and progress be monitored and instruction redirected in a responsive fashion?

Because of the disconnect between research and practice, teachers grapple with these questions for the most part without reference to a research base. The "what" and "how" might be answered by adherence to a textbook that reflects little of the research knowledge on student learning. And individual student differences in understanding are often overlooked entirely because textbooks rarely provide the tools for formative assessment and guidance on instructional responses to student difficulties. But as accumulating evidence points to the complexity of the underlying learning processes, the need for research-based knowledge to support professional practice becomes ever more apparent.

The decision to focus the agenda on practice has three closely linked entailments: (1) that the program of work be highly interdisciplinary, drawing on the range of knowledge bases and competences required to improve practice; (2) that the program emphasize and tightly integrate research and development and be carried through all stages necessary for classroom relevance; and (3) that the interdependence of student learning, teacher knowledge, and the organizational environment that characterizes practice be reflected in the SERP program of work.

THE RESEARCH BASE

To answer the questions that define practice will require an effort to bring together types of research and development that are now done separately, often by researchers who work in isolation from each other, and from program developers. If the questions of practice raised above are presented schematically, they might appear as in Table 1.1, where the questions are mapped for both student and teacher learning. The elements in the schema for educational practice are informed by several very different knowledge bases (see National Research Council, 2000, 2001c).

TABLE 1.1 Schematic Questions for Teaching and Learning

	Student Learner	Teacher as Learner
Destination	What should s/he know and/or be able to do (regarding the discipline or topic)?	What should s/he know about the discipline or topic? What should s/he know about student learning of, and the teaching of, the discipline or topic?
Point of departure	What are the typical preconceptions and informal understandings that students bring to the topic?	What are the teacher's existing understandings about the topic and about student learning?
Route	What is the expected progression of understanding, and what are the predictable points of difficulty or hurdles?	What are the typical pre-service and in-service learning trajectories and what difficulties are likely to be encountered?
Vehicle	What curriculum/pedagogy and classroom norms and practices facilitate learning?	What factors/experiences facilitate learning?
Checkpoints/course corrections	How can individual student progress be monitored and instructional activities matched to current understanding?	How can progress be monitored and instructional activities matched to current understanding?

- *What students should know or be able to do* in an area is informed by disciplinary expertise. It requires an understanding of the core concepts around which the disciplinary knowledge is organized, characteristic methods of reasoning and problem solving, and language and patterns of discourse. What to teach becomes a matter not only of the information and skills considered desirable for students to possess, but also of helping the student to build the conceptual framework that transforms information into understanding.

- *Knowledge of students' common understandings and pre-conceptions of a topic and the expected progression of student thinking* requires careful research on the typical trajectory of understanding. In part this research attempts to identify the nature and limits of children's changing cognitive abilities with age and instruction. And in part it attempts to uncover common understandings that can either support learning (the ability to halve or double relatively easily in mathematics) or undermine it (the common belief that heat and temperature are the same thing). Research findings suggest that students' everyday understandings are resilient, even after specific instruction to the contrary. That resilience highlights the importance of a carefully designed research program to inform and support teaching to achieve conceptual change. Research of this sort is generally done by cognitive scientists and education researchers, although the knowledge may emerge from the experience of expert teachers and the observation of exemplary practice.

- *The instructional interventions to move students along a learning path* constitute the core of what is generally considered education. These interventions may be designed by curriculum developers, teachers, or researchers. But regardless of the source, the contributors to skill development, knowledge acquisition, and conceptual change are themselves a research agenda, and the effectiveness of the instructional approach a matter for empirical testing.

- *General and discipline-specific norms and practices to support student learning* define the rules for interaction in the classroom. Learning takes place in classrooms that are themselves communities. Every community is distinguished by norms for work and interactions, ranging from when and how people collaborate to how they speak with one another. Some of those norms are general, rooted in an understanding of schools in a democratic society; others are discipline-specific, for what it means to do mathematics differs from what it means to do literary analysis, chemistry, or history. In all cases, the relationships between particular norms and the learning outcomes of interest are a matter for empirical investigation.

- *Assessing a student's progress* is the task of research and development on methods and systems of assessment. This knowledge base can quite naturally be developed and tested in the context of curriculum R&D, but it may also draw on more fundamental research—such as research on the nature and measurement of "comprehension." A key element of this work is linking assessment results to instructional responses (see Appendix A).

The parallel knowledge base for teacher learning would be similar in its disciplinary foundations and theoretical underpinnings. But the learning context, the learning goals, and the subject-matter content (particularly regarding pedagogical content knowledge) are different enough that the research base on teacher learning is quite distinct from (if overlapping with) that on student learning.

These various knowledge bases are not adequately developed to support teaching practice. Nor is the infrastructure in place to bring together people with the variety of competences required and to link their efforts in a program of work. SERP proposes to create that infrastructure, and the agenda we propose must build those linkages.

IMPROVING RESEARCH *AND* DEVELOPMENT

While much remains to be done to shore up the knowledge base, it is also the case that existing relevant knowledge is rarely incorporated extensively into classroom teaching. That this is so is disappointing, but hardly surprising when one considers the complexity of the task with which teachers are confronted. They face a class of students with different needs, behaviors, and preparation for learning. They must make choices about how to provide appropriate and powerful learning experiences for their students. They must simultaneously attend to their students' understanding while planning next steps. They must manage and monitor the students' learning and their learning environment. A program that is coherent—that guides instruction from one class to the next—can be of tremendous help in managing complexity. While the findings of researchers may be *relevant* to practice, if they are not easily incorporated into the teacher's instructional program, they may not be *useful*. Indeed, they may simply add to the complexity of an already highly complex task.

More targeted and instruction-relevant research (Hiebert et al., 2002) would be a good place to start. We can take as an example the research on the misconceptions students harbor in physics. In the course of everyday experience, people develop understandings or models of how the physical world works: as one moves closer to a heat source, temperature rises. One then draws inferences based on one's experiences that are very often scientifically incorrect: in the summer, the earth is closer to the sun. A persuasive body of evidence suggests that the models of physical principles that students deduce incorrectly from everyday experience are powerful and resilient (National Research Council, 2000; Vosniadou and Brewer, 1989; diSessa, 1982). While students may "learn" physics in the classroom and even perform quite well on tests, outside the classroom they revert to their untrained model (see Box 1.1).

The principle at work in physics is at play in all disciplines, undermining the effectiveness of the educational process. Nu-

• •

BOX 1.1 Misconceptions About Momentum

Andrea diSessa (1982) conducted a study in which he compared the performance of college physics students at a top technological university with the performance of elementary schoolchildren on a task involving momentum. He instructed both sets of students to play a computerized game that required them to apply a force (using a job stick) to a simulated object moving across the computer screen (a dynaturtle) so that it would hit a target, and do so with minimum speed at impact. Participants were introduced to the game and given a hands-on trial that allowed them to apply a few taps with a wooden mallet to a ball on a table before they began.

DiSessa found that both groups of students failed miserably at the task. Given the momentum of the dynaturtle, students should have applied a light force very early to ensure minimum speed at impact. Despite their training, college physics majors, just like the elementary school children, applied the force when the object was just below the target, failing to take momentum into account. Further investigation with one college student revealed that she knew the relevant physical properties and formulas and would have performed well on a written exam. Yet in the context of the game, she fell back on her untrained conceptions of how the physical world works.

merous examples have been identified in history and mathematics, as well as in science (National Research Council, forthcoming). Such "relevant" knowledge regarding the tenacity of everyday understandings becomes usable for a teacher only when it is applied to the subject and topic that are being taught, and when it is incorporated into instructional activities that draw out and effectively work with students' preconceptions. These activities must have potential for being incorporated into existing instructional practice, or be an acceptable replacement for that practice, if they are to have an effect. Only the most extraordinary teachers will be able to undertake such a task on their own. Furthermore, only the most inefficient profession would require the design of individual solutions to a general problem.

To effectively "bridge" research and practice, then, a research and development program must generate and draw on existing robust, relevant knowledge from a variety of disciplines, elaborate that knowledge so that it is usable in instructional practice, and then incorporate it into carefully tested tools and programs, directed both at student learning and at teacher learning. The research and development must be closely intertwined, so that program features are designed in response to research knowledge (derived either from disciplinary research or from the study of educational practice), and so that knowledge is continuously revised in the iterative cycles of design, study, and redesign. This necessarily means that the research and practice relationship is neither linear nor unidirectional. Instead, researchers and practitioners must interact in meaningful, progressively more sophisticated ways. Research is not neatly packaged and sent out to teachers to be implemented. Instead, researchers and teachers are mutually engaged in research and development in the context of practice.

Critical to the notion of follow-through is that when research findings are compelling, sustained attention is required to ensure independent replication of research and evaluation results in the range of environments of intended use for an educational intervention. Because education is a complex enterprise in which any outcome is influenced by a variety of factors, the conditions that support success in one setting may not be understood until it is attempted in other settings in which conditions differ. Moreover, evaluations are often conducted only by the designers of the intervention or their "critical friends."

While many research findings are criticized for this reason, research attention in the field of education is rarely directed at independent replication. Too often promising outcomes mark the end point of a research endeavor, rather than the beginning of a research effort aimed at replication and scientific generalization.

Finally, to be broadly useful to practice, attention must be sustained through careful study of the scaling up of successful interventions. When researchers design and study educational interventions, the process of study itself entails involvement by the researchers. The tacit and explicit knowledge of the researcher may support the teacher's effective use of the intervention. Similarly, an expert teacher's tacit knowledge and skill may be crucial to an intervention's success. But when the original researcher and teachers are no longer present, the innovation is often much harder to implement and sustain. Thus, the requirements for effective, sustainable use of an innovation must themselves be the subject of study.

INTERDEPENDENCE: STUDENT LEARNING, TEACHER KNOWLEDGE, AND ORGANIZATIONAL ENVIRONMENT

Whether in teaching or in medicine, practice is not embodied solely in the tools and protocols of the trade. Rather, these work in tandem with both the practitioner's knowledge and skill and the organizational environment. One shudders to imagine a magnetic resonance imaging (MRI) machine provided to doctors in a hospital with no more than an instruction manual on how to use it. Patients trust that doctors will be thoroughly versed in an understanding of the diagnostic power of the tool, the meaning that can be drawn from its images, and the confidence one can have in the information it provides under various circumstances. One also trusts that the organizational structure in which the physician operates will both appropriately support and impose standards on its use.

Similarly, in education, professional practice relies on all three simultaneously and interdependently: the tools and protocols designed for student learning, practitioner knowledge, and organizational structures. They serve as the three legs of the stool supporting student learning. While each is, in a sense, independent of the other two, the effectiveness of any one in supporting student achievement depends on the strength of the other two. It is a lesson learned repeatedly in education reform efforts: focusing on the strength of one without attending to the other two is a strategy that holds little promise for success. The SERP focus on practice, then, requires coordination not only of the knowledge from different fields on effective instructional programs to promote student learning, but also research on the knowledge requirements for teachers to carry out instruction effectively, and requirements of the environment to support effective instruction.

Teacher knowledge and skill matters a great deal in student learning. Ferguson (1991) analyzed data from 900 Texas school districts and found that teacher licensing exam scores, master's degrees, and experience accounted for over 40 percent of the variance in students' reading and math achievement scores after controlling for socioeconomic status. Other studies suggest a similarly powerful effect (Ferguson and Ladd, 1996; Strauss and Sawyer, 1986). Yet despite its importance, the research base on teacher learning is relatively undeveloped. SERP proposes to strengthen that knowledge base and, importantly, to tightly link the R&D on teacher learning and knowledge to that on student learning.

As we have already noted, the schematic questions that frame student learning also apply to teacher learning (see Table 1.1). For teachers, however, the questions that shape their practice provide a good start on our first question: "What should teachers know and be able to do?" The answer implied by the questions of practice is that teachers should understand the learning process of the student well enough to assess and guide it; understand the content to be taught well enough to select and use appropriate instructional materials; guide the pace and direction of instruction and flexibly respond to student questions and thoughts; understand the curriculum materials well enough to use them flexibly as a means to an end rather than as the end itself; understand the effects of classroom norms and practices well enough to create a supportive learning environment; and understand assessments well enough to interpret the outcomes and respond appropriately.

What is not well defined are the forms of knowledge a

teacher must master in order to reach that end and the levels of mastery needed. What mathematics must a teacher know, and what pedagogical knowledge does she or he need to make and implement appropriate decisions about the next best instructional steps to develop student thinking about gravity, for example? Although these questions are central to effective practice, little research has been done to provide answers.

Moreover, learning is as complex an undertaking when the teacher is the target as it is when the student is the target. A teacher's existing conceptions of teaching and learning, of student thinking, and of the subject matter must be understood and engaged. And experiences that bring about conceptual change for the teacher must be designed and effectively deployed for learning to occur.

The task is challenging; conceptual change is difficult to achieve when everyday experiences reinforce a misconception. In many everyday experiences one can simply tell people what they want or need to know. This is likely to influence a teacher's view about learning and instruction. When the frame of reference is common, simply telling works just fine. But when conceptions differ, telling is unlikely to be enough. "Elephants are bigger than pigs" may be completely adequate to communicate intended meaning, while "the orientation of the earth's axis relative to the sun determines the seasons" may be entirely inadequate. If the conception of teaching as telling is to be replaced by a more variegated model, powerful experiences that facilitate conceptual change on the part of the teacher about how students learn will be required.

The typical learning trajectory for teachers, and how it changes with learning opportunities, also requires empirical investigation. Much that teachers need to know cannot be learned apart from practice. This raises several questions for inquiry: Under what conditions can teachers best learn while engaged in practice? What knowledge and skill must teachers acquire at the beginning of their careers? What knowledge and skill is best acquired once they enter the profession? What organizational, material, and human resources are necessary to support and sustain teacher learning over time?

Widespread adoption and use of improved instructional methods are often hampered by institutional barriers that prevent or frustrate efforts to change: these may include problems of organizational structure, incentive structures, organizational culture, career patterns of teachers and administrators, and financial constraints. Even when presented with demonstrably effective instructional reforms, school systems are frequently incapable of moving the organizational machinery to achieve systemwide adoption (Briars and Resnick, 2000). And even when change is effectively instituted with the help of program developers, it is frequently not sustained when the developer departs.

Schools are certainly not unique in resisting change. Change is effortful. It imposes uncertainty and requires risk of failure. Resistance to change would be expected to weaken if uncertainty is reduced by providing effective supports for success, if risk is minimized by careful evaluation of candidate changes and the circumstances under which they are successful, and if effort and risk taking are rewarded. However change is not, in and of itself, a desirable end. There are certainly changes for the worse, and one hopes schools have mechanisms in place to resist such changes. What is desirable is an organization that can systematically assess its own performance, evaluate the potential of alternative approaches to improve performance, monitor the effect of change, and alter course to improve outcomes—an organization that can learn.

In the field of business management, attention has been devoted in the past few decades to the features of learning organizations. The focus of concern is corporations, and learning refers to the ability to incorporate new knowledge and technology required for effective competition and changing products to align with, or create, market demand. Business schools have drawn from a variety of disciplines: economics, statistics, political science, behavioral psychology, organizational psychology, and others. The potential contribution of these disciplines to the organization of medicine, agriculture, and the military has also been considered. However, their attention has not as yet been turned in a sustained way to the organization of schools.

If advances in instructional programs and teacher knowledge are to have a sustained impact on student learning, the organizational structure of schools must support that change. Because the success of each component (instructional program, teacher knowledge, and organizational structure) in contributing to improving student learning depends on the success of the others, all three must be integrated in a SERP research program.

The work of our panel, however, was to focus on learning and instruction. The committee that proposed a design for the SERP organization considers organizational issues to be of critical importance, as do we, and argues for a network on schools as organizations as a companion to the learning and instruction network from the start. To avoid mounting a research program that could otherwise be narrow and insular, it is essential that these two networks be closely tied as intended.

FRAMEWORK FOR A RESEARCH AGENDA

The committee considered two common strategies for organizing the research agenda: one focuses on specific subject areas taught in schools (science, history, etc.), and the other highlights research questions that cross subject domains (integrated assessment, teacher education, etc.). Much of the work on cross-domain issues is relevant regardless of the subject. This would include, for example, research on the relative effectiveness of professional development tools (like videotaped demonstrations, small group lesson study, etc.).

The proposed agenda embraces both, embedding the cross-cutting issues in subject-matter research. The rationale for this choice stems from the overarching commitment to focus on practice. We have argued that research is often not used in practice because it is not elaborated at the level of classroom practice, and classroom practice is subject specific. Furthermore, many of the cross-cutting issues are best illuminated with subject-specific examples. While the productive role that high-quality assessments can play in supporting effective instructional practice crosses topics, a deep understanding of the issues can be seen clearly in looking at a specific case, like the Force Concept Inventory described in Chapter 4.

The organization described in the *Strategic Education Research Partnership* is one in which regular stock-taking and coordination across research domains is given a high priority. In the panel's view, this coordination will be central to maximizing the potential of the program of research and development. Throughout the chapters that follow, the parallels across subjects are striking, and much can be learned if those engaged in different subject areas inform each others' work. Indeed, a good deal of the value added by a SERP organization is the opportunity created for the accumulation of knowledge, and the coordination of research protocols and data collection efforts across subjects that will make that accumulation possible.

Among the many cross-cutting issues, assessment deserves specific mention at the outset. The development of effective measures of the outcomes of learning and instruction is critical to producing high-quality evidence that can support the work of practitioners, policy makers, and researchers. Without an initial investment in developing reliable measures, even good-quality research leaves critical issues unresolved. Different assessments may show different instructional approaches to be beneficial, and often none of the assessment instruments is designed to fully capture the range of competences that are the desired outcome of instruction. Appendix A describes the nature of the work that in the committee's view needs to be done in the area of assessment research, development, and testing.

In the agenda described below, then, both cross-cutting and subject-specific topics are integrated into a program of research designed to support educational practice in the specific subjects targeted. The schematic questions posed in Table 1.1 provide a starting point for evaluating what is known currently that can support effective practice and where the research and development foundation is weak. We portray learning as a journey, using a travel metaphor to illuminate certain features of the schematic questions. Planning the trip requires first that one is clear about the destination, although it may change for a variety of reasons. There are multiple routes to get from a departure point to a destination, but routes will differ in the opportunities they afford for interesting experiences along the way, as well as in their efficiency at reaching the endpoint. Yet to reach the destination, the directional options are constrained, and there are likely to be critical points (passes) through which the path must lead. As with the route, there is no single vehicle required for a journey, but some work far better than others over a particular terrain. Finally, the challenge of monitoring progress is very different for different journeys. When a route is well marked, it can be a simple effort to clock the miles; but when the area is poorly mapped, frequent assessment and course correction are critical. For any subject matter taught, one can assess the quality of the research base by the knowledge it provides to support effective decision making regarding each of these issues.

In the long run, providing research-based knowledge to support answers to the schematic questions for every subject taught in schools is a desirable end, just as one expects the treatment by a physician of any ailment to be based on research-based knowledge. Yet the reality of the limited resources devoted to education research, as well as the existing capacity to conduct that research, suggest the need for focus on a limited set of subjects in order to ensure that work can be carried through all stages necessary for usability. As a knowledge base is consolidated in some areas, attention can be devoted to new areas.

CRITERIA FOR CHOOSING TOPICS

The areas in which research and development could be useful to educational practice are innumerable. To narrow the focus, the panel asked two questions:

1. Are there examples of rigorous research and development efforts that already show impressive gains in student achievement in research trials? These would provide the starting point for R&D that is intended as one of the hallmarks of SERP: following promising program outcomes with research on the circumstances under which the results are obtained, the feasibility of the intervention in the classroom and the school context, the teacher knowledge and support required for success, and the organizational factors that influence outcomes. We refer to these as "downstream" cases because the work in these areas has already traveled some distance toward classroom usability.

2. Are there pervasive problems of practice that are widely recognized as critical, but for which the knowledge base is too weak to guide instructional interventions? We refer to these as the "upstream" cases, since the work is still in early stages. Since the upstream research will need to go through all phases of research and development required for usability, the impact on practice is likely to require more time.

While question 1 is driven by the opportunities provided by

the current knowledge base, question 2 is driven by the needs of practitioners. Precisely because the knowledge base is weak for the latter, R&D on these questions would draw more heavily on the study of practice and the extension of research from the social and behavioral sciences to strengthen the education knowledge base.

As we worked through agenda development, the upstream-downstream distinction became somewhat blurred. The downstream examples often involve unanswered questions about teacher learning or taking practice to scale that are likely to require a sustained research effort. And in areas in which the knowledge base is not well consolidated, there are individual practices or curricula that appear promising. Nonetheless, the distinction continued to be useful as an indicator of the potential impact on practice in the short run.

RESEARCH DOMAINS

The panel recommends three areas for focus: reading, mathematics, and science. These three domains emerged when we looked for downstream areas in which current research and development shows promising results that could substantially influence practice. In both mathematics and reading, the strongest research bases address learning in the early elementary years. In science, in contrast, physics was identified as the area that is furthest downstream—a subject generally taught in high school.

All three domains are recommended for upstream research as well, but for reasons that are not uniform. Practitioners on the panel emphasized the central role standardized tests play in the lives of students, particularly as they reach the upper grades. There was considerable disagreement regarding the desirability of these tests and a common concern regarding the ability of tests to distort and undermine good instructional practice. But the tests play a role in setting the emotional stage for students, touching their sense of identity and self-confidence. Since performance in mathematics and in reading comprehension is a major contributor to test results, these are seen as important areas for improving both student outcomes and the experience of schooling itself. The choice of these areas was further supported by the need for proficiency in order to successfully meet the demands of modern life.

Reading comprehension also features prominently in the teaching of most subject areas, including English, history and social studies, mathematics, and science. Teachers of these subjects must decide to what extent reading comprehension instruction is required as part of their effort to teach the subject matter. There is currently little guidance in this regard. Research on reading comprehension therefore has the potential to provide benefits in virtually all subject areas. Furthermore, reading comprehension is poorly measured. Current tests emphasize the speed of reading and short-term recall of factual information. But existing research suggests that both speed and short-term recall are weak predictors of the construction of understanding from text that comprehension requires. Given the importance of reading comprehension in standardized tests, making progress on test measurement issues would have substantial potential to influence practice.

In contrast to mathematics and reading comprehension, science was identified for upstream research because the development of science curricula, particularly for the elementary and middle school years, has been remarkably weak. The American Association for the Advancement of Science recently reviewed widely used textbooks in middle school science. "The study probed beyond the usual superficial alignment by topic heading and examined each text's quality of instruction aimed specifically at the key ideas, using criteria drawn from the best available research about how students learn" (Roseman et al., 1999). Not one of the middle school science texts evaluated by the project was rated as satisfactory. High school biology texts scored slightly higher than the middle school texts, but the evaluation found serious shortcomings in both their content coverage and instructional design (Budiansky, 2001). Textbooks across the grades were characterized as "overstuffed and undernourished," with presentation of a great many facts and too few opportunities to present the concepts that make those facts meaningful (Budiansky, 2001). In contrast to mathematics, there is little agreement in science as to the sequence and content of study, or even when science education should begin.

The areas the committee has chosen for focus are strategic: they provide either the opportunity to leverage existing investments in research by carrying promising findings through to practice, or they hold promise for providing new knowledge in areas of critical need. We emphasize, however, that a SERP network would probably undertake a more thorough investigation of candidate opportunities, and it is likely to be shaped in significant measure by the opportunities that emerge for productive work in schools as SERP is launched. In particular, the link between cognition and context is likely to be better understood and incorporated into the developing agenda as the SERP work progresses.

· ·

ORGANIZATION OF THE REPORT

In describing our proposed agenda for research and development, we organize the work by domain. Chapters 2, 3, and 4 address reading, mathematics, and science in sequence. Each discipline is treated at some length because our purpose is to consider the program of research and development that would be required to strengthen the knowledge base on the entire set of questions that define teaching practice, as well as an understanding of the knowledge requirement for teachers of the subject. Because we focus so broadly, however, we do not go into depth in any single area. Readers looking for more depth are referred to other reviews intended for that purpose.

Structuring the program around the questions of practice as we propose brings coherence to the agenda, but it may tax the reader who finds the same issues of assessment or teacher knowledge treated similarly in three different chapters. We consider the repetition in the agenda important, however, because the state of the art in education research and development is one in which many of the most critical principles of instruction are understood at a general level, but the work of R&D to incorporate those principles into the teaching of individual subjects has not been done. That work must be repeated for each subject if learning and instruction are to be improved.

At a time when the quality of education research is a matter of heated debate not only among researchers, but also in the halls of Congress, careful consideration of methodological rigor and quality control are critical elements of a research agenda. The final chapter discusses how the proposed agenda addresses contemporary concerns about education research quality and impact. We also consider the extent to which proposed SERP features are required to effectively carry out that agenda.

2 Reading

Mastering the mechanics of fluent reading and the ability to comprehend text provides a key to acquiring knowledge in all other domains of learning. Given the escalating demands for text comprehension that pervade virtually every aspect of contemporary life in the United States, as well as the stagnating performance of students in reading mastery and reading comprehension, strengthening the knowledge base in ways that can support practice should be a high priority.

This chapter is divided into two parts: one on early reading and one on reading comprehension beyond the early years. In many respects the parts overlap: teacher education is a major theme in both, as is research and development to improve instructional interventions. In the school context, reading is treated differently in the early years than it is in the years after third grade. From fourth grade on, the emphasis switches from learning to read to reading to learn. Because a central purpose of the SERP work is to provide R&D that is useful to practice, we adopt the division used by schools. We would expect the SERP R&D network to operate in a highly interactive fashion, with research and methods used on one set of questions informing those used on other, related research efforts.

EARLY READING

STUDENT KNOWLEDGE

The Destination: What Should Children Know and Be Able to Do?

There is an unusual degree of consensus regarding the goals of early reading instruction. The consensus is captured in the National Research Council report, *Preventing Reading Difficulties in Young Children* (National Research Council, 1998) and in the report of the National Reading Panel, *Teaching Children to Read* (National Institute of Child Health and Human Development, 2000). The goals are often expressed in terms of the competencies children should be able to demonstrate at the end of third grade: (a) reading age-appropriate literature independently with pleasure and interest, (b) reading age-appropriate explanatory texts with comprehension for the purpose of learning, and (c) talking and writing about those texts in age-appropriate ways. Achieving these goals requires simultaneous development of an interdependent set of abilities: decoding skills, reading fluency, oral language development, vocabulary development, comprehension skills, and the ability to encode speech into writing.

The Route: Progression of Understanding

The foundation for early reading lies in the earlier, informal acquisition of language. With little effort, children with intact neurological systems acquire the sounds of their language, its vocabulary, and its methods of conveying meaning (National Research Council, 1998). The path that children travel in acquiring language is predictable (National Research Council, 1998), although the age at which particular skills and abilities are mastered varies somewhat. Babies comprehend words during the first year of life generally well before they can produce them. Once production begins, usually during the second year, vocabulary grows steadily, and single word utterances become sentences that increase in length and complexity. As the ability to produce and understand more complex sentences develops, children are less reliant on the immediate context to support meaning. This "decontextualized" language eases the transition to school, where it is the common parlance.

As proficiency with language use grows, children develop the ability to *think about* language. Before that ability develops, they do not distinguish between the word and the object to which it refers. "Snake" is thus deemed to be a "long" word, and "caterpillar" a "short" word (National Research Council, 1998:50). Children can begin to develop rudimentary meta-linguistic skills as early as age 3. Acquiring this ability allows children to play with, analyze, and pass judgment on the correctness of language.

The trajectory of language development described above is universal, although the richness of the environment affects the pace and extent of language development powerfully (Hart and Risley, 1995; Huttenlocher, 1998). For example, Graves and Slater (1987) found that first graders from higher income families had a vocabulary that was double the size of those from lower income families. The differences are highly relevant because verbal ability generally, and vocabulary development particularly, are good predictors of success in early reading.

While normal language development supports reading acquisition, other abilities required for effective reading mastery are unlikely to develop unless children receive formal instruction. With few exceptions, children need systematic instruction in the alphabetic principle to learn to decode words and to learn how to encode words in writing (Adams et al., 1998). This instruction is what is referred to as "phonics." But successful phonics instruction rests on a more fundamental ability: phonemic awareness. This is the awareness, for example, that the word "cat" consists of three separable sounds: c/ a/ t. The distinction is important because phonics instruction that teaches the mapping of separate sounds onto letters requires for success that a student hear those separate sounds.

Learning the alphabetic principle is a prerequisite to reading. However, it is not nearly sufficient to help children reach the desired third grade competencies. Phonics instruction must be integrated with comprehension instruction, opportunities to develop fluency in reading through practice, instruction to enhance and practice oral and written language abilities, and opportunities to acquire rich vocabulary and background knowledge. The failure of any one of these will result in falling short of the third grade goals. If fluency does not develop, little meaning is taken from a text that a child must plod through. If background knowledge is inadequate, even a fluent reader will be unable to engage with and learn from the text.

The components of successful reading are tightly intertwined. There is strong empirical support, for example, for the relationship between young children's oral language and subsequent reading proficiency (Bishop and Adams, 1990; Scarborough, 1989; Share et al., 1984). Oral language, vocabulary in particular, is essential for understanding the text that is read. In addition recent longitudinal research suggests that young children's vocabulary is associated with improved decoding skills (Lonigan et al., 2000; Wagner et al., 1997), as well as growth in phonological sensitivity (Bowey, 1994; Lonigan et al., 1998, 2000; Wagner et al., 1993, 1997).

In addition to building vocabulary, oral language instruction can extend a child's ability to understand and use academic, or literate, language. This is the decontextualized language that minimizes contextual cues and shared assumptions (e.g., by explicitly encoding referents for pronouns, actions, and locations; Olson, 1977; see Box 2.1).[1] These extensions of discourse in the decontextualized register of academic language are what predict literacy success into middle school, controlling for home variables (Dickinson and Sprague, 2001). These relationships between preschool oral language and middle school reading comprehension are clearly mediated by decoding instruction in the primary grades (Whitehurst and Lonigan, 2001). But the point is that language intervention that builds vocabulary and decontextualized language structures needs to occur *prior* to and *during* decoding instruction, rather than later.

Writing is at the heart of mastering the alphabetic system. Writing starts with the encoding of speech to print. The ability to phonemically segment sounds in speech and represent them in conventional writing develops over time. A complete repre-

[1]Decontextualized language is at the core of literacy instruction because it allows literate individuals to communicate without personal interaction. This literate language or academic language is a specific oral language register valued in traditional schooling (Cook-Gumperz, 1973; Heath, 1983). For example, in conversation we rely on shared context to disambiguate pronouns and referents in the sentence "He wants to go there next time." In written language we are expected to clarify the who, where, when (e.g., *Sally wants to go to the beach the next time you make a trip.*)

sentation of a word's spelling in memory developed through writing will enhance the speed and accuracy with which it is recognized (Ehri, 1998; Perfetti, 1992). Thus, the writing of words supports the reading of words and, over time, builds toward the writing of text, which can support the comprehension of text.

In addition to understanding the contributors to successful reading acquisition, there is also an extensive research base on the typical hurdles that children encounter (National Research Council, 1998; National Institute of Child Health and Human Development, 2000). It is now well established that a significant number of children have difficulty learning the alphabetic principle because they have not developed phonemic awareness. Among children who learn to decode words but do not comprehend well, fluency is often the culprit; if children struggle slowly through a text, their comprehension when they have finished will be poor. Fluency can suffer if children spend too little time actively engaged in effective reading practice, or if vocabulary and background knowledge are too weak to allow the student to read with understanding.

. . . .

BOX 2.1 Decontextualized Language Instruction in the Early Years

Links between decontextualized language and literacy have been made by Dickinson and his colleagues in a longitudinal study of 85 children from low-income families started in 1987 (Dickinson and Tabors, 2001; Dickinson and Sprague, 2001). Significant *prekindergarten* variables that influenced literacy development were quality of teachers' talk and curriculum quality. Quality of teacher talk was measured in rare word usage, ability to listen to children and to extend their comments, and tendency to engage children in cognitively challenging conversations (i.e., conversations about nonpresent topics). The prekindergarten variables of quality of teacher talk, vocabulary environment, and curriculum quality predicted kindergarten outcomes above and beyond home variables, thereby emphasizing the importance of instruction in literate language and a quality emergent literacy curriculum in preschool classrooms for children from low-income homes.

Composite variables that significantly influenced *kindergarten* literacy and vocabulary scores were home variables of literacy support, density of rare words used, and extended discourse. Kindergarten outcomes in turn predicted vocabulary and reading comprehension scores in middle school (Dickinson and Sprague, 2001).

The Vehicle: Curriculum and Pedagogy

Because there is general agreement on the goals of early reading instruction and a fairly solid research base on the contributors to success in achieving those goals, the targets of quality early reading instruction are fairly clear: phonemic awareness and phonemic decoding skills, fluency in word recognition and text processing, construction of meaning, vocabulary, spelling, and writing (Foorman and Torgesen, 2001). Box 2.2 reproduces the curricular components for first to third grade recommended by the National Research Council Committee on the Prevention of Reading Difficulties in Young Children.

Knowing the components of effective instruction is critical, but it is only a start. It falls far short of knowing how to effectively integrate the various components, how to allocate time among those components, and how to carry out instruction effectively in a classroom context in which children differ significantly in their preparation for learning and rate of progress on each of the components. And when a goal of instruction is to motivate children to read for pleasure, the *how* of instruction is at least as important as the *what*.

The research literature makes clear that there is no single answer to the instructional questions posed (National Research Council, 1998; Fletcher et al., 2002). Effective reading teachers use a variety of instructional strategies and curricula successfully. For example, while teaching phonics is a critical component of effective instruction, there are multiple approaches to doing so effectively (see Box 2.3). Many teachers rely largely on basal readers for teaching the alphabetic principle, supplemented by trade books for children's reading practice, guided reading for comprehension instruction, and books read aloud for vocabulary development and further comprehension modeling. Other teachers place more emphasis on children's writing as one source of instruction about the alphabetic principle, systematic minilessons in grapheme-phoneme correspondences using word sorts and other procedures (but no basals), and careful assignment of trade books to children for practice and comprehension instruction. Some reading curricula that have shown gains for groups of at-risk children involve fairly strict adherence to instructional scripts to deliver all the literacy instruction to highly homogeneous reading groups, whereas others use

BOX 2.2 Curriculum Components for Reading Instruction in First Through Third Grade

Beginning readers need explicit instruction and practice that lead to an appreciation that spoken words are made up of smaller units of sounds, familiarity with spelling-sound correspondences and common spelling conventions and their use in identifying printed words, "sight" recognition of frequent words, and independent reading, including reading aloud. Fluency should be promoted through practice with a wide variety of well-written and engaging texts at the child's own comfortable reading level.

Children who have started to read independently, typically second graders and above, should be encouraged to sound out and confirm the identities of visually unfamiliar words they encounter in the course of reading meaningful texts, recognizing words primarily through attention to their letter-sound relationships. Although context and pictures can be used as a tool to monitor word recognition, children should not be taught to use them to substitute for information provided by the letters in the word.

Because the ability to obtain meaning from print depends so strongly on the development of word recognition accuracy and reading fluency, both of the latter should be regularly assessed in the classroom, permitting timely and effective instructional response when difficulty or delay is apparent.

Beginning in the earliest grades, instruction should promote comprehension by actively building linguistic and conceptual knowledge in a rich variety of domains, as well as through direct instruction about such comprehension strategies as summarizing the main idea, predicting events and outcomes of upcoming text, drawing inferences, and monitoring for coherence and misunderstandings. This instruction can take place while adults read to students or when students read themselves.

heterogeneous grouping for most instruction, reverting to homogeneous grouping only to teach specific skills.

To extend the metaphor, there are different vehicles that are capable of making the journey to the desired destination. But the route traveled by any vehicle must cover certain territory to reach the destination, including decoding territory, oral language and vocabulary development territory, comprehension territory, and writing territory. But while there are multiple instructional approaches used by effective teachers, there are many ineffective teachers whose students do not reach the destination at all. The National Assessment of Educational Progress (NAEP) for the year 2000 found 37 percent of all fourth graders and 60 percent of black and Hispanic graders reading below the

Once children learn some letters, they should be encouraged to write them, to use them to begin writing words or parts of words, and to use words to begin writing sentences. Instruction should be designed with the understanding that the use of invented spelling is not in conflict with teaching correct spelling. Beginning writing with invented spelling can be helpful for developing understanding of the identity and segmentation of speech sounds and sound-spelling relationships. Conventionally correct spelling should be developed through focused instruction and practice. Primary grade children should be expected to spell previously studied words and spelling patterns correctly in their final writing products. Writing should take place regularly and frequently to encourage children to become more comfortable and familiar with it.

Throughout the early grades, time, materials, and resources should be provided with two goals: (a) to support daily independent reading of texts selected to be of particular interest for the individual student and beneath the individual student's frustration level, in order to consolidate the student's capacity for independent reading, and (b) to support daily assisted or supported reading and rereading of texts that are slightly more difficult in wording or in linguistic, rhetorical, or conceptual structure in order to promote advances in the student's capabilities.

Throughout the early grades, schools should promote independent reading outside school by such means as daily at-home reading assignments and expectations, summer reading lists, encouraging parent involvement, and working with community groups, including public librarians, who share this goal.

SOURCE: National Research Council (1998).

"basic" level (National Center for Education Statistics, 2001).[2] It is therefore of utmost importance that knowledge gleaned from the study of effective practice and from research on reading instruction be carefully articulated, tested in rigorous research, and incorporated into instructional programs and into teacher education programs.

Checkpoints: Assessment

The general consensus in the field of early reading that has emerged from a relatively strong research base extends to the

---

[2]Categories include below basic, basic, proficient, and advanced.

early indicators of reading difficulties. Effective indicators are emerging from longitudinal databases (e.g., Fletcher et al., 2002; O'Connor and Jenkins, 1999; Scarborough, 1998; Torgesen, 2002; Vellutino et al., 2000; Wood et al., 2001). These indicators can provide valuable information to teachers regarding the instructional needs of individual students. Predictiveness of particular skills depends on how and when they are assessed, but, in general, phonological awareness and its theoretically related construct of letter-sound knowledge in kindergarten and the beginning of first grade are predictive of first grade outcomes, as is word recognition at the beginning of first grade. In second grade word reading continues to be a strong predictor of second grade outcomes, with reading fluency and reading comprehension becoming increasingly important predictors of reading outcomes. For children at risk of reading difficulties due to poverty and language background, oral language in general and vocabulary in particular are critical to reading success (Foorman et al., in press; National Research Council, 1998; Dickinson and Tabors, 2001).

There are a number of models for screening children in kindergarten, first grade, and second grade for reading prob-

· ·

BOX 2.3   Phonics Instructional Approaches

Analogy phonics: teaching students unfamiliar words by analogy to known words (e.g., recognizing that the rime segment of an unfamiliar word is identical to that of a familiar word, and then blending the known rime with the new word onset, such as reading <u>brick</u> by recognizing that <u>ick</u> is contained in the known word <u>kick</u>, or reading <u>stump</u> by analogy to <u>jump</u>).

Analytic phonics: teaching students to analyze letter-sound relations in previously learned words to avoid pronouncing sounds in isolation.

Embedded phonics: teaching students phonics skills by embedding phonics instruction in text reading, a more implicit approach that relies to some extent on incidental learning.

Phonics through spelling: teaching students to segment words into phonemes and to select letters for those phonemes (i.e., teaching students to spell words phonemically).

Synthetic phonics: teaching students to explicitly convert letters into sounds (phonemes) and then blend the sounds to form recognizable words.

SOURCE: National Institute of Child Health and Human Development (2000).

lems.[3] These assessments engage teachers—some more and some less formally—in collecting data on which to base curricular decisions about individual children. Both the Texas Primary Reading Inventory (TPRI) and the Virginia Phonological Awareness and Literacy Screening (PALS) have been implemented on a statewide basis.

Many children who master the process of reading nonetheless do poorly at developing broader literacy skills (RAND, 2002a). While there are fairly good predictors of difficulty in learning to read, predictors of comprehension problems are less well developed. Assessments of vocabulary and writing ability, both of which support comprehension, are underdeveloped. While there are many standardized tests of vocabulary and writing in use, they provide only normative information relative to students in the same age or grade and therefore are not adequate to provide individual feedback that can guide instruction. Vocabulary tests that assess the breadth and depth of word meanings are required to give teachers information about which words to target for instruction. Likewise, writing protocols that are evaluated for spelling, mechanics, grammar, word choice, ideas, and organization are needed to provide the basis for the revision process so fundamental to the development of skilled writing.

TEACHER KNOWLEDGE

Reading teachers need to understand the current state of knowledge on the course of literacy development, and the role of reading instruction in supporting that development. The specific areas of study that would align teacher preparation with the learning experiences that should be provided to children in the classroom are outlined in detail in *Preventing Reading Difficulties* (National Research Council, 1998:285-287). We are far from the goal of effectively providing all reading teachers with

---

[3]These include the Observation Survey developed in New Zealand (Clay, 1993); the South Brunswick, New Jersey, Early Literacy Portfolio (Salinger and Chittenden, 1994); the Primary Language Record (Barr et al., 1988); the Work Sampling System (Meisels, 1996-1997); the Texas Primary Reading Inventory; and the Phonological Awareness and Literacy Screening developed at the University of Virginia (see Foorman et al., 2001, for summaries of all of these programs).

access to that knowledge base. But we know a great deal about what the knowledge is. There are two major teaching challenges, however, that are likely to require more than understanding the knowledge base: integrating the components of early reading instruction into an effective reading program and differentiating instruction for children with different competences.

Integrating Instruction Knowing the components of effective reading instruction does not ensure that a teacher will be able to integrate these in practice. With the multiple demands of managing a classroom, teachers often look to curriculum materials to simplify their complex task. For the majority of teachers in the United States, these are basal readers (National Research Council, 1998). The content of the basals influence's how teachers allocate instructional time. The research base on phonics instruction is stronger than that on vocabulary instruction, oral language instruction, writing, and comprehension. Perhaps because dimensions of effective phonics instruction are better defined, basal reading programs place more emphasis on phonics. In many places, considerably more time is now invested in phonological awareness teaching and practice than would be recommended based on the research, which suggests that 18-20 hours is sufficient (National Institute of Child Health and Human Development, 2000).

With little attention paid to vocabulary and writing instruction in the basals, Foorman and Schatschneider (in press) observed little attention to these critical activities (i.e., less than 10 percent of instruction) in observations of 114 first and second grade teachers in 17 high-poverty schools. The National Reading Panel concluded that "teachers must understand that. . . . systematic phonics instruction should be integrated with other reading instruction in phonemic awareness, fluency, and comprehension strategies to create a complete reading program" (National Institute of Child Health and Human Development, 2000:11). But this is not now standard practice.

Differentiating Instruction Children who have mastered the alphabetic principle simply do not need phonics instruction that is as intense as that needed by their peers who have not achieved such mastery. Furthermore, as Box 2.2 suggests, a critical aspect of an effective instructional program is daily practice reading within the student's mastery level and daily supported practice working with text that is just beyond mastery level. With children at multiple mastery levels in a classroom, an effective teacher must be able to assess current mastery, provide students with work that is appropriate, and manage a classroom in which groups of students are engaged in different activities.

Without the opportunity for teachers to learn how to integrate research knowledge into instructional practice, the knowledge has a weak influence at best. Most basal reading instruction, for example, happens at the whole-class level. Students are rarely grouped for instruction because of concerns about managing the rest of the class and lack of knowledge about how to translate assessment results into small-group instruction. Most pre-service teacher credential programs do not provide coursework on assessment, and most reading basals do not provide assessments that clearly translate to differentiated instruction. Exceptions are well-implemented Success for All and Reading Mastery programs. The result is a preponderance of whole-class instruction in which, if assessment is present at all, results are not linked to instruction tailored to the needs of individual students.

RESEARCH AGENDA

The research base supporting principles of good literacy instruction has been in place for at least 20 years, and policies and strategies designed to improve reading outcomes have similarly been a priority at the federal level and in many states for a long time. Nonetheless, many children in U.S. schools are not learning to read well, and in many primary classrooms good teaching practices are not being implemented. While accountability and incentives may be part of the solution, the committee believes there are important gaps in knowledge and know-how that must be filled if effective reading instruction is to become the norm. The discussion above points to three critical questions:

1. How can we ensure that all children have the foundational experiences to support success in reading mastery? How can oral language development, including vocabulary acquisition, be supported in the preschool and early elementary years to counter the

effects of disadvantage? The question is particularly urgent for the growing numbers of children whose native language is not English.

2. How can the knowledge base on components of effective reading instruction be refined to address important instructional questions and classroom management issues? The goal must be to move from *principles* of good literacy instruction to the *practices and programs* that will support it.

3. How can pre-service and in-service teacher education be designed and combined to most effectively support the development of expertise in teaching reading?

These are the questions that motivate our early reading R&D initiatives.

Initiative 1: Narrowing the Gap

Awareness of the critical role of early experiences in preparing children for school success has been heightened in recent years by data collection efforts that document striking differences among socioeconomic groups when children first pass through the schoolhouse door (West et al., 2001). Several recent NRC reports have emphasized the importance of addressing these disparities in the preschool years.[4] The report on preventing reading difficulties argues that preschool and kindergarten programs should pay ample attention to the skills that play a causal role in future reading achievement (National Research Council, 1998:9):

> Instruction should be designed to stimulate verbal interaction; to enrich children's vocabularies; to encourage talk about books; to provide practice with the sound structure of words; to develop knowledge about print, including the production and recognition of letters; and to generate familiarity with the basic purposes and mechanisms of reading.

Federal policy makers have responded with concern, as the Reading Excellence Act, the Early Reading First Guidelines, and

---

[4]See National Research Council (1998, 2001a, 2002a).

the No Child Left Behind legislation attest. With seeming agreement about what should be done, the time is ripe for investing in research and development on how to do it. How can preschool programs best enrich the oral language skills, including the vocabulary, of young children? How does the answer differ for children ages 3, 4, and 5?

There are examples of interventions for preschool and kindergarten programs that are designed to build children's capacities with practices that the current knowledge base suggests are important. For classroom purposes, these must be subjected to systematic evaluation to determine whether they are indeed effective—both in general and for subgroups of children (e.g., English-language learners). These practices include the following:

- Regular use of read-alouds that focus on engaging children in discussion of the text and that offer opportunities to reuse and to expand on the meaning of the more challenging vocabulary items in the text (e.g., Whitehurst et al., 1994; Lonigan and Whitehurst, 1998; Valdez-Menchaca and Whitehurst, 1992; Beck and McKeown, 2001).
- Use of science-, number-, or world-knowledge-focused curricula to raise the quality of talk going on in the classroom and thus on children's language growth (for examples of such curricula, see National Research Council, 2001b).
- Increasing the amount of one-on-one or small group, adult-child conversation during the daily activities of the classroom, since considerable evidence (e.g., Dickinson and Tabors, 2001) suggests that such opportunities are both relatively rare and highly facilitative of children's language growth.
- Professional development programs that provide rich practice-embedded knowledge about vocabulary and oral language development, whether or not paired with explicit guidelines about the use of such activities as dialogic reading, text-talk sessions, science curricula, and so on.

Initiative 2: Models of Integrated Reading Instruction

Moving from the *principles* to the *practices* of effective reading instruction will require attention to the detailed instructional decisions teachers must make on a daily basis. How much attention should be paid to writing, to oral language development, and to decoding? At the same time that the National Reading Panel emphasized the importance of what research says about phonics instruction, for example, they point to the instructional questions that remain unknowns (National Institute of Child Health and Human Development, 2000:10):

> If phonics has been systematically taught in kindergarten and 1st grade, should it continue to be emphasized in 2nd grade and beyond? How long should single instruction sessions last? How much ground should be covered in a program? Specifically, how many letter-sound relations should be taught, and how many different ways of using these relations to read and write words should be practiced for the benefits of phonics to be maximized?

Phonics is by no means atypical with regard to unanswered instructional questions. Reading practice is generally recognized as an important contributor to fluency. Two instructional approaches—guided repeated oral reading and independent silent reading—are generally used to support fluency development. While evidence regarding the effectiveness of the first is strong, evidence on independent silent reading is more mixed (National Institute of Child Health and Human Development, 2000). Either approach, however, uses valuable instructional time. Questions regarding the amount of time that is optimal for reading practice at each stage of the learning process and whether the combination of oral and silent reading that is desirable changes as mastery level changes are central to everyday instruction. Similarly, while much is known about the importance of vocabulary to success in reading, there is little research on the best methods or combinations of methods of vocabulary instruction.

To make headway on the instructional questions, the research and development we propose would consist of two closely sequenced and intertwined efforts. The first is the study of exemplary teaching practice in early reading instruction, and the second is the design and study of specific interventions, with particular attention to basal readers.

Learning from Exemplary Practice We know that very successful reading teachers manage to integrate the components of early reading instruction. But many teachers are less successful. An important resource for advancing understanding of the practices that promote student success is the study of contrasting practices and outcomes.

One approach to this research would be to identify teachers who consistently beat the odds with the performance of their students on the full range of literacy skills. Observing these teachers—the constellation of practices that they employ, the mix of activities, the distribution of time spent on various tasks, and the assessment measures to which they attend and respond—would allow for hypothesis formation regarding the features of effective, integrated reading instruction programs. These beat-the-odds teachers would be compared with teachers in the same school and teaching the same grade level whose students consistently make only average progress for their school, in order to identify the crucial features that differentiate the two groups of teachers.

These features are likely to differ by grade, and by the average achievement level and language development of the students in the classroom. Designing the research to look at various levels (kindergarten, first, second, third grade) and at classrooms chosen to represent a wide variety of demographic factors (e.g., in suburban high-scoring schools, in urban low-scoring schools, in schools serving language minority learners) would be required to draw implications for practice.

A first level of analysis of the data would be to see whether the features that differentiate the beat-the-odds teachers are the same for different groups of learners. It is entirely plausible that the characteristics of excellent teaching for inner-city or for English-language learners are different from those that work best with suburban youth. It is even more strongly to be suspected that the characteristics of excellent instruction in one year (e.g., third grade) will vary from the instruction that has gone on in earlier years. Clarity about these and other dimensions that require instructional responsiveness is an important target for this research.

Developing and Testing Reading Intervention Once hypotheses are generated regarding the components of effective instruction for different groups and grades, the next phase of R&D would involve the design of interventions that incorporate those components into instruction systematically, in an effort to verify their effectiveness experimentally and to assess their efficacy with a wider array of students and reading curricula or subject matter.

Since basal readers are the primary reading curriculum materials for the majority of American classrooms (National Research Council, 1998), they are an obvious target for improvement in reading instruction. Two studies have examined first grade basal readers for their theories of learning (Foorman et al., 2002; Hiebert et al., 2002), and both studies report their limitations. For example, the vast majority of words presented in text selections in a lesson in first grade basals are used only once, yet research clearly indicates that multiple presentations of a word are required before it becomes part of a student's vocabulary (National Institute of Child Health and Human Development, 2000). Basals also differ significantly on the decodability of the text (the composition with respect to length, grammatical complexity, the number of unique and total words, repetition of words, and coverage of important vocabulary). Iterative cycles of design and research on the features of instructional interventions generally, and basals specifically, could make a direct contribution to instructional practice.

The development projects should be undertaken in competing efforts in order to maximize creativity and entrepreneurship, and each project should be conducted with a research component to test critical features (e.g., variation in time spent on vocabulary instruction.) Once instructional interventions have gone through sufficient iterations of design, testing, and redesign, the interventions should be tested more broadly. More powerful, large-scale, longitudinal, planned variation studies could be undertaken to test the relative benefit of different instructional programs, and data on student achievement results should be collected to ascertain how large an impact the intervention has on students with different characteristics (socioeconomic status, primary language, achievement level, etc.). Simultaneously, teacher knowledge and support requirements should be studied.

Initiative 3: Teacher Education

While the research base is quite strong on the elements of effective reading instruction (what we want teachers to be able to do), we know surprisingly little about how to provide teachers with the learning experiences that can support effective practice in teaching reading (National Institute of Child Health and Human Development, 2000).

We do know, however, that existing teacher preparation has not been adequate to support widespread use of research-based practices in the classroom (Moats and Lyon, 1996; Moats, 1994). The problem is as difficult as it is important. Teachers have long-standing beliefs about student learning and teaching practice that are built on personal experience, and many believe that a knowledge base in pedagogy is not needed (Lanier and Little, 1986; RAND, 2002a). Some research indicates that even teachers who say they use reform models use traditional practices (Stigler et al., 1999). Heibert and Martin (2001) found, for example, that teachers distort knowledge about mathematics reform to make it consistent with what they already do. True changes in practice are difficult to effect. Yet mounting evidence suggests that the quality of teaching strongly predicts student achievement, explaining as much as 43 percent of the variance after controlling for socioeconomic variables (Ferguson, 1991; National Research Council, 2002a). Clearly, if student learning is the ultimate goal, teacher learning must be a target. In developing the agenda, we ask what research and development would support more effective learning for teachers of early reading.

The first question a research team must address is how teacher learning, and the effect of teacher learning on student learning, will be measured. The problem is not trivial. Key to success is a teacher's ability to effectively integrate and differentiate instruction. These are multifaceted, complex phenomena. How they can be captured in measurable dimensions of teacher practice will itself require careful research. And the impact of teacher instruction on student achievement is equally complex, relying as it does on multiple literacy skills, including the difficult-to-measure variable of "comprehension."

Once the outcomes of interest are identified, the question for research is how to achieve them. A report by RAND (2002a) asks the question for reading comprehension that could just as easily be asked of all teacher education for early reading: "What is the relative power of various instructional delivery systems (e.g., field-based experiences, video-based cases, demonstration teaching, microteaching) for helping teachers acquire the knowledge and skills they need to successfully teach comprehension to students of different ages and in different contexts?" (p. 51). In the fields of medicine and business, in which much of what needs to be learned by practicing professionals is how to draw effectively on a knowledge base to support good judgment, education programs generally place heavy emphasis on case-based learning. Similarly, the ability to move successfully from study to practice, from knowing *"what"* to knowing *"how,"* is likely to require supervised practice (internships) and mentoring relationships. Indeed, there is evidence to suggest that coaching and mentoring models that translate research into instructional activities in actual classrooms are more likely to have an impact on teacher development and hence student learning (Bos et al., 1999; Foorman and Moats, in press; McCutchen et al., 2002; O'Connor, 1999; Thomas et al., 1998). The effectiveness of various approaches and the optimal combinations of approaches to teaching teachers are empirical question's of such wide-ranging importance that there can be little debate that they deserve research attention.

The specific questions that might be the subject of research on teacher education, however, are innumerable. Consider those posed by the National Reading Panel: "What is the optimal combination of preservice and inservice education, and what are the effects of preservice experience on inservice performance? What is the appropriate length of inservice and preservice education? What are the best ways to assess the effectiveness of teacher education and professional development? How can teachers optimally be supported over the long term to ensure sustained implementation of new methods and sustain student achievement gains? The relationship between the development of standards and teacher education is also an important gap in current knowledge" (National Institute of Child Health and Human Development, 2000:17). The long-term agenda must be one in which this broad set of questions is addressed in an attempt to make the massive system of teacher education more rational and effective. The benefit of a SERP network is that these questions can be taken on sequentially over time.

How might the network set priorities for early focus? We can apply the previously stated principle that existing, important research findings should be carried through to practice. The very powerful research base discussed above suggests that many of the students who do not learn to read could be effectively instructed if teachers could assess individual student needs and differentiate instruction in response to the assessments. A high priority, then, might be given to teacher learning in this area. What do teachers need to know to grasp the goals of individual assessments, to administer and score the assessments competently, and to understand their implications? What do they need to know to effectively group students and align assessment results with instruction? And how can teachers be prepared for the demands of managing a classroom in which different groups of students are working on different tasks, some independently and some with teacher guidance?

A key advantage of the network is that it can monitor progress on research questions and steer the agenda over time. As progress is made on one set of questions, the next set can follow. Even more importantly, knowledge from research on teacher learning with respect to early reading instruction can be integrated in the network context with research on teacher learning with respect to mathematics and science.

• •

READING COMPREHENSION BEYOND THE EARLY YEARS

STUDENT KNOWLEDGE

In the area of reading comprehension, the panel had the benefit of drawing on a thorough and very recent assessment of research needs by the RAND Reading Study Group (RAND, 2002a), as well as on the report of the National Reading Panel (National Institute of Child Health and Human Development, 2000). Those reports make clear that with regard to both student learning and teacher preparation, the research base to support practice is weak.

The Destination: What Should Children Know and Be Able to Do?

An answer to this question is implied by the RAND study group in its definition of reading comprehension as "the process of simultaneously extracting and constructing meaning through interaction and involvement with written language" (RAND, 2002a:11). To *extract* meaning requires the reader to decode the words and form a mental representation of what the text actually says, at both a local (sentences, phrases, and their interconnections) and global level (the gist of the text's meaning). To *construct* meaning requires that the reader create a "situation model," or an understanding of the intended meaning conveyed with these words that is informed not just by the text, but also by the knowledge and experience of the reader (Kintsch, 1998). The situation model is the foundation from which inferences are drawn (see Box 2.4). Consider the sentence, "The sky was a clear, bright blue the day she first saw Charles." The sentence does not state that it is not raining, but the reader can infer this from the bright blue sky. More importantly, it says nothing about who Charles might be to the referenced woman, but we infer that he will be significant and memorable—not a plumber who will fix her drain then disappear.

· ·

BOX 2.4 Text Comprehension Involves Processing at Different Levels

First, there is the linguistic level, the **text** itself. The reader must decode the graphic symbols on a page. Perceptual processes are involved, as well as word recognition and parsing (the assignment of words to their roles in sentences and phrases).

Semantic analysis determines the **meaning of the text**. Word meanings must be combined in ways stipulated by the text, forming idea units or propositions. However, there is more to the meaning of a text than word meanings and propositions. The global structure of a text is often crucial for comprehension. Psychologically, these processes involve the determination of the coherence relations among the propositions expressed in a text (which are often, but not always, signaled by linguistic markers). Inferences, such as simple bridging inferences or pronoun identification, are often necessary. Macrostructures require the recognition of global topics and their interrelationships, which are frequently conventionalized according to familiar rhetorical schemata.

But if a reader comprehends only what is explicitly expressed in a text, comprehension will be shallow, sufficient perhaps to reproduce the text, but not for deeper understanding. For that, the text must be used to construct a **situation model**, that is, a mental model of the situation described by the text. Generally, this requires the integration of information provided by the text with relevant prior knowledge, as well as the goals of the reader. One important fact to note about the process of constructing situation models is that it is not restricted to the verbal domain. It frequently involves imagery and emotions, as well as personal experiences.

The Route: Progression of Understanding

We would be pleased if a 6-year-old student could read the above sentence and understand it semantically. But we would expect a 16-year-old student to develop a situation model that is more complex due to greater developmental maturity, more experience with texts and text genres, and the benefits of instruction. The high school student might appreciate the expectation created by the author with two very simple phrases and might productively reflect on how that expectation might change if the sky were dark and the wind threatened to carry away all in its path. And yet understanding of the typical progression of student reading comprehension between ages 6 and 16 is poorly mapped, with a consequence that instructional support for comprehension is poorly defined as well. As the RAND study group argues, "without research-based benchmarks defining adequate progress in comprehension, we as a society risk aiming far too low in our expectations for student learning."

Research in this area is, for the most part, still upstream. Many research perspectives offer relevant insights (Pearson and Hamm, 2002), but as yet there are no integrated theories and companion models that provide a foundation for accumulating knowledge and guiding instruction. Moreover, mapping progress in reading comprehension requires that the phenomenon can be measured. Here again the knowledge base is weak. Worse, what is known suggests that existing, commonly used measures of comprehension can be misleading. They capture meaning extraction and short-term memory, but these are not good predictors of meaning construction. Interventions that can improve short-term recall can actually weaken inferencing capacity (Mannes and Kintsch, 1987). Both the mapping of progress in reading comprehension and the evaluation of instructional interventions to improve reading comprehension depend on the development of assessments that can measure all its aspects, including the quality of the situation model.

The Vehicle: Curriculum and Pedagogy

Instruction in reading comprehension is generally limited. Research in the 1970s indicated that only 2 percent of classroom reading instruction time was devoted to comprehension of the text being read (Durkin, 1978-1979). In the various domains that place heavy demands on reading comprehension abilities, like social studies, science, and even mathematics, little or no comprehension instruction takes place. Yet mastering the vocabulary, text structures, methods, and perspectives of the discipline places simultaneous demands on the student to acquire content knowledge and reading skill. While teachers in a variety of domains recognize that for many students subject-matter knowledge is held hostage to reading comprehension skill, they are given no preparation or guidance in providing reading comprehension instruction.

Many teachers assume that children will acquire comprehension skills simply by reading. While some do, many do not. Research done 25 years ago found that many readers of various ages who were given text with logical and semantic inconsistencies failed to detect them (Markman, 1977, 1981; National Institute of Child Health and Human Development, 2000). The surprising failure of comprehension focused attention on the complexity of the comprehension process, as well as on the nature of the activity as one involving active engagement rather than passive reception (National Institute of Child Health and Human Development, 2000). Research on comprehension instruction in the decades since has focused on strategies to actively engage the reader.

To the extent that reading comprehension is addressed at all in the K-12 curriculum, it is generally done through *strategy instruction* (RAND, 2002a). Common approaches to strategy instruction focus on strengthening recall of textual materials. Study skills instruction, for example, typically teaches students skills like previewing texts, paying attention to headings, rereading for specific information or structural cues, outlining or mapping the text in graphic form, and rehearsing questions to prepare for a test (Kintsch et al., 2001). These strategies have been shown to improve recall, particularly for low-achieving students (National Institute of Child Health and Human Development, 2000). But because they focus on surface features, they can be mastered successfully without the student's developing a situation model or integrating the new knowledge with the student's background knowledge.

A number of programs were developed to help combat the surface-level reading identified as a widespread problem (Brown et al., 1983). The programs focus on the development of metacognition, a term that refers to the ability to monitor and guide one's own thinking processes. Students who are good comprehenders actively monitor whether their purpose in reading is being met. They notice when something is unclear or is inconsistent with their background knowledge. Metacognitive strategy instruction teaches students to consciously use problem-solving strategies to comprehend difficult text, to activate relevant background knowledge, and to stay alert to comprehension breakdowns. Several have demonstrated promising outcomes.

Reciprocal teaching, a program developed by Palincsar and Brown (1984), is based on the notion that the internal thought processes involved in effective comprehension can be taught to students explicitly. The teacher initially models aloud four strategies: questioning unclear content, summarizing meaning paragraph by paragraph, clarifying comprehension problems, and predicting what will come next. Students practice the strategies under the teacher's guidance, and gradually the teacher's role diminishes. As students become more adept, they take on the role of the teacher themselves in small "cooperative learning" reading groups, asking their own questions aloud. Over time, students internalize the comprehension process. Reciprocal teaching can be used to support listening comprehension among younger children, as well as reading comprehension once children become fluent readers.

Questioning the author is an instructional strategy that supports deeper comprehension by changing the nature of the questions students are asked about the text they read. Rather than focusing on factual questions that direct the student to retrieve information from text, Beck et al. (1997) found that focusing on interpretation and intent became a powerful tool for changing students' approach to reading comprehension. In questioning the author, students are asked to think about the author's intended message and to evaluate how successfully that message is conveyed. The strategy calls attention to gaps in understanding and stimulates retrieval of existing knowledge against which to judge the author's case. Beck et al. (1997) have documented impressive changes in the classroom culture, with students more actively engaged in interactive discourse. The dynamic is self-reinforcing; as students and teachers engage more in thoughtful questioning, they become more critical readers and thinkers. A version of the approach, called text talk, was developed to support listening comprehension in young children.

Transferring strategy use effectively from a particular classroom context to other classrooms or to contexts outside the school has proven difficult (National Research Council, 2000; RAND, 2002a). Several quasi-experimental studies suggest the benefit of embedding strategy instruction in content learning (Guthrie et al., 1998a, b). The purpose of the strategy as a tool for understanding challenging text and the need to adapt the strategy to the activity are both more apparent when embedded in content. Students become more proficient at deliberate strategy use as a means for learning (Brown, 1997).

For well over a decade, research findings have confirmed the benefits of metacognitive strategy instruction (Pressley et al., 1989; Rosenshine and Meister, 1994; Rosenshine et al., 1996). In reviewing the studies that met their methodological standards, the National Reading Panel concluded, "when readers are given cognitive strategy instruction, they make significant gains on measures of reading comprehension over students trained with conventional instruction procedures" (National Institute of Child Health and Human Development, 2000:4-40).

Research also suggests that the classroom norms and practices that promote reading comprehension are those that enhance student motivation and engagement (RAND, 2002a). That motivation is important is hardly surprising given that reading comprehension requires effortful engagement of multiple cognitive processes. The RAND review concludes that teachers who give students choices, challenging tasks, and opportunities for collaborative learning increase their motivation to comprehend text. Both classroom observation (Turner, 1995) and quasi-experimental studies (Reeve et al., 1999) suggest the importance of student choice and autonomy that are limited but meaningful in increasing motivation.

Checkpoints: Assessment

Unlike word decoding, comprehension is not an isolated ability the mastery of which can be straightforwardly measured. Comprehension takes place at the intersection of *the reader*, *the text*, and *the activity*. Assessing progress of an individual student and diagnosing problems requires attention to that interaction. While the person must bring or develop the requisite skills, what is requisite will depend on the text to be comprehended and the purpose for which it is being read.

The Reader The capacity to comprehend text varies enormously across students. Some of the contributors to that variation are well established (see National Institute of Child Health and Human Development, 2000, and RAND, 2002a, for elaboration). These include the following:

- Comprehension capacity builds on successful initial reading instruction;
- Comprehension capacity is coincident with good oral language skills (vocabulary and listening comprehension);
- Students who have had rich exposure to literacy experiences are more likely to succeed at reading comprehension;
- Maturing cognitive capacities, including attention, memory, analytical ability, inferencing ability, and visualization ability, contribute to comprehension ability;
- Good comprehenders actively monitor their understanding and use strategies that help them retain, organize, and evaluate information;
- The growing store of background knowledge acquired both inside and outside school contributes to comprehension capacity; and
- The motivation that the student brings to reading contributes to comprehension.

The Text Features of a text can facilitate or complicate comprehension. But the relationship is not simple. We know how to make a text easy or hard to read, for example, by controlling vocabulary and syntax, by using an appropriate rhetorical structure, and by calibrating the information in the text to the readers' prior knowledge. However, what is crucial for comprehension is neither the reader nor the text alone, but the reader-text interaction. The goal is to engage the reader as actively as possible, so that the reader will utilize his or her prior knowledge to make inferences and construct a situation model that integrates this prior knowledge with the newly acquired textual information. A text designed to be maximally readable may leave the reader passive, resulting in the readers' knowing the textbook by heart (and doing very well on an exam that tests for rote knowledge). But that knowledge may be inert, inflexible knowledge that will not help the student to solve novel problems and understand new texts requiring that the knowledge be brought to bear.

The problem, then, requires careful attention. If the text is too hard and relies on background knowledge that the reader does not have, he or she will not be able to construct a good situation model and comprehension will fail. If it is too easy, comprehension may remain superficial.

The Activity The pedagogical problem for comprehension is to get the reader to engage in the right kind of processing, in which what is "right" depends on the purpose of the activity.

Different purposes for engaging with text place very different demands on the reader and suggest different standards for adequate comprehension. Skimming a text for particular information that can help solve a problem is quite different from studying a text to be conversant with its entire content. Reading a novel for pleasure imposes different demands from reading the same novel in order to critique features of the author's technique and style. The same reader may have good comprehension skills for one purpose and weak skills for another.

Assessing progress and making course corrections both require focus on all three factors. While strengthening a particular weakness of the reader may be the appropriate response to inadequate progress in some cases, providing a text that is more appropriate to the reader's background knowledge or providing activities that have a greater capacity to engage the student may be more appropriate in others. The teacher's task is complex. Assessments that can assist a teacher in diagnosing the nature of the problem for an individual student have not yet been developed.

TEACHER KNOWLEDGE

What do we want teachers to know and be able to do with respect to reading comprehension? The answer proposed in the RAND report is to "enact practices that reflect the orchestration of knowledge about readers, texts, purposeful activity, and contexts for the purpose of advancing students' thoughtful, competent, and motivated reading"(RAND, 2002a:29-30). That orchestration is highly complex because the amount of support that individual students need is likely to differ considerably, and because the support any given student needs may change with different texts. This kind of "adaptive expertise" requires that a teacher have a deep understanding of comprehension processes and approaches to supporting them. Mastering decontextualized rules of instruction will not be adequate (National Research Council, 2000).

Metacognitive strategy instruction is a case in point. To teach metacognitive strategies effectively, teachers must have a good grasp themselves of the text content and of the strategies. But their knowledge must be conditionalized; they must know which strategies are most effective for which students and which types of content. They must be able to respond flexibly and opportunistically when the intervention is needed to aid student comprehension rather than using strategy instruction as the end in itself.

Research by Palincsar et al. (1989) suggests that the preconceptions that many teachers hold regarding student learning are at odds with the research-based conceptions incorporated in reciprocal teaching. In one study with first grade teachers, for example, the teachers' own goals for student listening comprehension emphasized the ability to follow a sequence of directions, and instruction was limited to the support of that goal. The teachers believed that collaboratively constructing the meaning of text was beyond their students' abilities. Pilot studies of reciprocal teaching using teachers whose beliefs were in accordance with the program showed significant gains for 85 percent of students. But in the hands of first grade teachers with discordant beliefs, only 47 percent of their first graders showed comparable improvement, even after the teachers were trained in the technique by the researchers.

In other work as well, Palincsar and her colleagues found that teachers who conceptualize reading as a sequence of isolated skills require considerable support and coaching to overcome the propensity to use the strategies in a routine fashion. Without that support, their students show gains in tests of strategy use, but not in reading comprehension (Palincsar, 1986). Teachers' conceptions regarding collaborative learning more generally diverge from the research base as well. While research suggests that student performance in collaborative groups exceeds individual performance, teachers generally believe that collaborative groups support social goals but not cognitive gains (Palincsar et al., 1989).

The National Reading Panel reviewed quantitative studies of teacher preparation in the area of strategy instruction for reading comprehension and found four that met their methodological criteria. The conclusions drawn from these few studies reinforce the findings above, suggesting that to be effective teachers required extensive instruction in explaining what they were teaching, modeling their thinking processes, encouraging student inquiry, and keeping students engaged. But the studies provided little guidance on which aspects of the teacher preparation were most effective (National Institute of Child Health and Human Development, 2000).

RESEARCH AGENDA

In the area of reading comprehension, the panel proposes four initiatives:

1. to conduct R&D on assessments of reading comprehension;
2. to conduct R&D on the teacher knowledge requirements for effective use of proven approaches to metacognitive strategy instruction;
3. to advance the knowledge base on the components of effective comprehension instruction across grades; and
4. to undertake a benchmarking effort to define expectations for comprehension across the school years.

Initiative 1: Assessments of Comprehension

All research on the effectiveness of interventions to support reading comprehension requires that the phenomenon of comprehension be measured. In laboratory research, different levels of text comprehension can be distinguished (Kintsch, 1998; Graesser et al., 1997). Comprehension can be deep in the sense that the information is integrated with prior knowledge and can be used for problem solving or other purposes. In contrast, comprehension can be shallow, focusing on the text itself. In this case, new knowledge is inert; it is not integrated into the reader's general knowledge base or used when applicable for problem solving.

Current reading comprehension assessments do not effectively capture this distinction. Tests that are widely used today require students to answer questions that often assess only superficial aspects of comprehension. If deeper understanding of text is the goal of instruction, those tests will be inadequate to inform decisions about instructional effectiveness. Indeed, Box 2.5 suggests that inferencing ability and recall are different aspects of comprehension, and measuring recall only can be highly misleading. New assessments must therefore be a high priority.

Since comprehension is not a unitary process, it is necessary to assess separately the different components of comprehension. Just how many independent components exist is a matter of some disagreement (Pearson and Hamm, 2002). Rigorous research to push further on a working answer to that question is under way and should be continued and extended. Hannon and Daneman (2001), for example, designed a comprehensive test that measures four different components of comprehension: a reader's ability to recall a text, the ability to make inferences based on explicitly stated facts, the ability to access general word knowledge, and the ability to make inferences that require integration of prior knowledge with text information. These four components proved to be good predictors of performance on a variety of comprehension tasks. Kintsch et al. (2001) pursued a similar goal when they assessed separately how well people could reproduce a text and how well they could answer simple problem-solving (inference) tasks for which information in that text is required.

Importantly, promising initial developments must be followed through to ensure both their validity and their practicality. Evaluation of internal construct validity examines how well the assessment explains comprehension performance in com-

. . . .

BOX 2.5 Measuring Recall Alone Does Not Measure Comprehension

In a study by Mannes and Kintsch (1987), students read one of two versions of a chapter: one was well organized and explicit; the other was slightly disorganized and left some things unsaid. When asked to recall the chapter, the well-organized version produced 25 percent more recall. However, when understanding was tested by inference questions, the less explicit version was better by 75 percent. Making readers draw their own inferences when studying had its benefits, but if measured by a test that merely required them to reproduce the text, the reverse would appear true.

parison to what is considered a reliable measure of that performance. The benchmark might be the ratings on one or more in-depth interviews that thoroughly explore and rate students' comprehension of a text. Questions of practicality must also be investigated. Measuring comprehension could be made more reliable if testing time and scoring time were not constraints. However both are valuable resources, and balancing quality and practicality will require attention. Box 2.6 gives an example from the recently revised SAT. It provides a measure of deep comprehension in the very practical (for scoring purposes) multiple-choice format.

But perhaps most importantly for improving educational outcomes, the instructional validity of the assessment must be investigated: Does it provide information that can be used to productively shape an understanding of the student's instructional needs? Can it help guide the teacher's instructional decisions? The SAT question posed in Box 2.6 provides insight into the student's ability to understand the literary use of a word in context that requires a fairly sophisticated understanding. But

• •

BOX 2.6 Comprehending Text on the Revised SAT

In its recent revision of the SAT, the College Board includes the item below in the verbal section:

> Dinosaurs have such a powerful grip on the public consciousness that it is easy to forget just how recently scientists became aware of them. A 2-year-old child today may be able to rattle off three dinosaur names, but in 1824, there was only one known dinosaur. Period. The word "dinosaur" didn't even exist in 1841. Indeed, in those early years, the world was baffled by the discovery of these absurdly enormous reptiles.

The statement "Period" in the middle of the paragraph primarily serves to emphasize the:

(A) authoritative nature of a finding
(B) lack of flexibility in a popular theory
(C) stubborn nature of a group of researchers
(D) limited knowledge about a subject
(E) refusal of the public to accept new discoveries

Answer: (D).

SOURCE: *Education Week* (2002).

the assessment information becomes useful for instruction only if instructional approaches to developing that understanding are known and available to teachers.

While we discuss this research first, clearly the ongoing work of developing and improving assessment tools must be directly influenced by other components of the research agenda. The definition of comprehension used to establish internal validity will depend on what one wants students to know and be able to do. As research knowledge improves, the benchmarks change (see Initiative 4), and so must the assessments. SERP will be a particularly productive environment for this iterative work, because the research on these interdependent questions is conducted within a single, well-integrated network.

Initiative 2: Teacher Knowledge and Metacognition Strategy Instruction

While reading comprehension has been defined as upstream because much of the fundamental work to understand and measure the phenomenon has not yet been done, strategy instruction provides an opportunity for improving comprehension that is further downstream. Several programs, including reciprocal teaching, text talk, and questioning the author, provide clearly articulated approaches to metacognitive strategy instruction that substantially improve reading comprehension, particularly for struggling students. While the techniques involved can be simply and briefly described, to effectively execute strategy instruction in the classroom proves a challenge for many teachers.

The panel therefore recommends an R&D effort to develop instructional materials, protocols, and supports for teachers who are learning to use strategy instruction in the classroom. The effort should include careful identification of teacher conceptions regarding student learning and effective instruction in reading comprehension, as well as their divergence with the research base that undergirds the intervention.[5] Instructional

---

[5]Clearly there is overlap across disciplines in research on teacher conceptions of student learning and of effective teaching. One benefit of carrying out a program of research in the context of a network on learning and instruction is that the work on various research projects within the network would be closely integrated and designed so that results can easily be compared and accumulated.

experiences that can support conceptual change should be developed and tested for effectiveness.

Because reading comprehension is a problem across all disciplines, the research should be conducted separately with teachers in different fields. The nature of the texts that allow teachers to master the nuances of matching strategy use with text type and comprehension goals should be studied carefully. As in the teacher learning initiative in early reading, a variety of tools for supporting teacher learning should be considered separately and in combination.

The development should be undertaken as a research effort, in which program components or tools are tried and tested with classroom teachers. Dimensions of variation to be studied should include discipline and grade level taught, experience level, and school demographic characteristics (e.g., suburban high-scoring schools, in urban low-scoring schools, schools serving language minority learners). The products should be tested during and after development for their impact on teacher understanding, changes in practice, and attitudes toward the product or program. Dimensions of effective use should also be examined, including the time and feedback required for mastery.

Initiative 3: Instructional Practices to Support Reading Comprehension

The variety of contributors to successful reading comprehension listed above suggest there are many potentially productive avenues for improving student performance. But from the perspective of practice, the list begs for greater clarity regarding which factors have the greatest capacity to influence comprehension outcomes, which can be most effectively influenced instructionally, and which interventions are most productive, in which combinations, at which ages. Indeed, these questions are so fundamental to practice in so critical an area that our current ignorance is rather astonishing.

Research could be designed to test the "which" questions in much the same way as research on effective reading instruction in earlier years. Teachers who consistently beat the odds with the performance of their students in the area of reading comprehension could be identified and compared with other teachers whose students consistently achieve less. The observation would include the constellation of practices that they employ, the mix of activities, the distribution of time spent on various tasks, and the assessment measures to which they attend and respond. This work must be coupled with research to test the core hypotheses experimentally, so that the causal mechanisms are clarified.

As with reading instruction in the earlier years, features are likely to differ by grade and by the average achievement level and language development of the students in the classroom. Designing the research to look at various levels (e.g., third, sixth, and ninth grade) and at classrooms chosen to represent a wide variety of demographic factors (e.g., in suburban high-scoring schools, urban low-scoring schools, schools serving language minority learners) would again be required to draw implications for practice with specific attention to differences for students in different demographic groups and in different grades.

The next phase of R&D would involve the design of interventions that incorporate those components into instruction systematically, in an effort to verify their effectiveness experimentally, and to assess their efficacy with a wider array of students and reading curricula or subject matter. The interventions should address both teacher learning and student learning.

The design and testing phases of the development projects would look much like those in the initiative on early reading interventions. An important benefit of an R&D network is that the expertise in doing this type of work begins to accumulate in an organizational setting in which what is learned—both in research outcomes and in the *conduct of research and development*—has a continuing influence on future R&D projects and designs.

Initiative 4: Benchmarks for Comprehension

Simultaneously with the efforts to explore, identify, then rigorously test best practices in comprehension instruction, it is crucial that the educational research and practice communities collaborate with stakeholders, such as future employers, faculty in community colleges and universities, and others interested in educational outcomes, to define what adequate reading comprehension is for readers of various ages. The SERP networks are in a unique position to engage in this kind of stock-taking on a regular basis. Indeed, this is an important raison d'être of the networks.

Ultimately, of course, defining what readers should be able to comprehend constitutes a decision based on a society's values and willingness to invest in education as much as it does on research. Research can make a unique contribution, however, regarding what is achievable at what ages. Examples include research on children's developing abilities to consider simultaneously opposing perspectives, to understand the motivations and intentions of others, and to distinguish between their own beliefs and knowledge. Furthermore, establishing reasonable benchmarks for comprehension can be informed by considering historical, international, and economic perspectives.

A useful place to begin is with an understanding of the endpoint. What should an 18-year-old be able to comprehend? One dimension that would enter into any benchmarking process is defined by the complexity of the text she or he might be expected to read. Presumably 18-year-olds should be expected to understand general-purpose texts—newspapers, magazines, novels, popularized presentations of science or history, introductory university level textbooks—with no difficulty. Reading these texts presupposes a minimum vocabulary of 40,000 words, the capacity to process fairly complex syntax, and some flexibility in processing various discourse structures.

But the readability or complexity of the text is only one dimension defining comprehension level. Another is the depth of processing that one can expect of the text being read. One might well expect an 18-year-old to understand nonliteral uses of language in text, for example, irony, parody, sarcasm; to appreciate stylistic niceties in text; to consider the possibility that "factual" texts include intentional misrepresentations, biased or limited perspectives, incomplete representations of reality, and errors; to process fiction as being about themes or issues and not just about plot; to appreciate cogent arguments in texts with which the reader nonetheless disagrees; and to read for attitude and perspective rather than just for information.

If the capacities outlined above can reasonably be expected of an 18-year-old reader, then what are the developmental benchmarks that characterize progress toward that point? Once the endpoint is established, it is relatively easy to work backward in order to define the expected comprehension capacity of younger students. If these are defined for the dimension of text complexity, then a linear projection is probably reasonable: children learn a certain number of new vocabulary words a year, and there is no strong reason to believe, for example, that during some years they can only learn fewer and during other years can learn more. But the depth-of-processing dimension is limited by children's cognitive capacities—their theory of mind level, their capacities to take others' perspectives, to coordinate multiple perspectives, and to distinguish belief from knowledge. We would need to call on the knowledge built up from basic research in cognitive development in order to establish reasonable expectations about children's capacities for deeper processing of texts. Elaborations of that work might well be helpful in deciding at what age the majority of children would be most susceptible to being taught about multiple perspectives in text, or about the use of textual features to raise doubts or questions, or other subtleties of processing.

Thus benchmarks for comprehension, while not themselves a simple matter of drawing conclusions from research findings, could be deeply informed by a set of research activities that considered both basic cognitive development and the array of standards identified by various groups with a wide array of interests and experiences. Establishing an initial set of benchmarks, and developing assessments of them, would help improve outcomes in reading comprehension simply by proffering a common understanding of what needs to be taught and learned. The initial set of benchmarks should be subject to regular review and recalibration. In fact, the comprehension-instruction agenda outlined above and the benchmarking agenda sketched here should be constantly confronting one another. As improved instruction in reading comprehension raises learners' capacities, the benchmarks can be ratcheted up to ensure that the proper balance between high standards and opportunities to succeed is maintained.

3 Mathematics

Debates about mathematics instruction have long focused on the relative importance of developing fluency with mathematical procedures and developing the ability to reason mathematically. Few on either side of the debate would disagree that both are necessary for competence in mathematics. There is disagreement, however, on the relative weight and share of instructional time to be given to each and on the approach to instruction that best supports mathematical competence.

Investment in recent decades by federal agencies and private foundations has produced a wealth of knowledge on the development of mathematical understanding and numerous curricula that incorporate that knowledge. As a result, elementary mathematics is ripe for investment in rigorous, independent evaluation to compare the outcomes of alternative approaches to teaching mathematics across a range of students, teachers, and contexts and making that knowledge usable and used widely by schools.

• •

ELEMENTARY MATHEMATICS

STUDENT LEARNING

The Destination: What Do We Want Children to Know or Be Able to Do?

U.S. students fare poorly in international comparisons of mathematics achievement. They show weak understanding of basic mathematical concepts, and although they can perform straightforward computational procedures, they are notably weak in applying mathematical skills to solve even simple problems (National Research Council, 2001c). These results have generally been attributed to the shallow and diffuse treatment of topics in elementary mathematics relative to other countries (National Research Council, 2001c).

The panel had the benefit of drawing on a recent synthesis of research on elementary mathematics (National Research Council, 2001c) and on the work of a RAND study group that produced a mathematics research agenda (RAND, 2002b). The National Research Council report presents a view of what elementary schoolchildren should know and be able to do in mathematics that draws on a solid research base in cognitive psychology and mathematics education. It includes mastery of procedures as a critical element of mathematics competence, but it places far more emphasis on conditional knowledge: understanding when and how to apply those procedures than is common in mathematics classrooms today. Conditional knowledge is rooted in a deeper understanding of mathematical concepts and a facility with mathematical reasoning. The NRC committee summarized its view in five intertwining strands that constitute mathematical proficiency (National Research Council, 2001c:5):

- *Conceptual understanding:* comprehension of mathematical concepts, operations, and relations;
- *Procedural fluency:* skill in carrying out procedures flexibly, accurately, efficiently, and appropriately;
- *Strategic competence:* ability to formulate, represent, and solve mathematical problems;
- *Adaptive reasoning:* capacity for logical thought, reflection, explanation, and justification;
- *Productive disposition:* habitual inclination to see mathematics as sensible, useful, and worthwhile, coupled with a belief in diligence and one's own efficacy.

A well-articulated portrait of mathematical proficiency is an important first step; it provides a well-defined goal for mathematics instruction. But important questions remain regarding the allocation of time and attention to the separate strands, as well as the approach to instruction that best supports the proficiency goal.

The Route: Progression of Understanding

Research has uncovered an awareness of number in infants shortly after birth. The ability to represent number and the development of informal strategies to solve number problems develop in children over time. Many studies have explored how preschoolers and children in the early elementary grades understand basic number concepts and begin operating with number informally well before formal instruction begins (Carey, 2001; Gelman, 1990; Gelman and Gallistel, 1978).

Children's understanding progresses from a global notion of a little or a lot to the ability to perform mental calculations with specific quantities (Griffin and Case, 1997; Gelman, 1967). Initially the quantities children can work with are small, and their methods are intuitive and concrete. In the early elementary grades, they proceed to methods that are more general (less problem dependent) and more abstract. Children display this progression from concrete to abstract in operations first with single-digit numbers, then with multidigit numbers. Importantly, the extent and the pace of development depend on experiences that support and extend the emerging abilities.

Researchers have identified two issues in *early* mathematics learning that pose considerable challenges for instruction:

1. Differences in children's experiences result in some children—primarily those from disadvantaged backgrounds—entering kindergarten as much as two years behind their peers in the development of number concepts (Griffin and Case, 1997).

2. Children's informal mathematical reasoning and emergent strategy development can serve as a powerful foundation for mathematics instruction. However, instruction that does not explore, build on, or connect with children's informal reasoning processes and approaches can have undesirable consequences. Children can learn to use more formal algorithms, but may apply them rigidly and sometimes inappropriately (see Boxes 3.1 and 3.2). Mathematical proficiency is lost because procedural fluency is divorced from the mastery of concepts and mathematical reasoning that give the procedures power.

BOX 3.1 Buggy Algorithms

When students attempt to apply conventional algorithms without conceptually grasping why and how the algorithm works, "bugs" are sometimes introduced. For example, teachers have long wrestled with the frequent difficulties that second and third graders have with multidigit subtraction in problems such as:

$$\begin{array}{r} 51 \\ -14 \\ \hline \end{array}$$

A common error is:

$$\begin{array}{r} 51 \\ -14 \\ \hline 43 \end{array}$$

The subtraction procedure above is a classic case: Children subtract "up" when subtracting "down"—tried first—is not possible. Here, students would try to subtract 4 from 1 and, seeing that they could not do this, would subtract 1 from 4 instead. These "buggy algorithms" are often both resilient and persistent. Consider how reasonable the above procedure is: in addition problems that look similar, children can add up or down and get a correct result either way:

$$\begin{array}{r} 51 \\ +14 \\ \hline 65 \end{array}$$

Bugs often remain undetected when teachers do not see the highly regular pattern in students' errors, responding to them more as though they were random miscalculations.

BOX 3.2 Rigid Application of Algorithms

Many examples can be cited in which students attempt to plug numbers into algorithms without thinking about their meaning, a phenomenon that stretches through all grades of schooling and all mathematical subjects. Even when students are capable of solving a problem correctly informally, they are found to produce incorrect answers when they use formal algorithms. In studies by Lochhead and Mestre (1988), for example, college students who were told that there are 6 times as many students as professors, and there are 10 professors, could correctly give the number of students. But when students are asked to write the formula to represent that situation, the majority write $6S = P$. The formula seems correct to students even though the solution would yield 6 times as many professors as students. The occurrence of the word 6 near the word students is sufficient to lead to a formal representation of the problem that is at odds with their informal knowledge.

The Vehicle: Curriculum Development

Past investments in R&D have produced curricular interventions to address each of the two problems raised above. With respect to the first, several curricula have been developed that introduce children to whole-number mathematics, with particular attention to the needs of young children who have had little preparation outside school. The most extensively researched of these is the Number Worlds curriculum, which has been tested in more than 20 matched, controlled trials. The results suggest that well-planned activities designed to put each step required in mastering the concept of quantity securely in place can allow disadvantaged students to catch up to their more advantaged peers right at the start of formal schooling (see Box 3.3). The curriculum has a companion assessment tool (the Number Knowledge Test) to help the teacher monitor and guide instruction. If results in controlled trials could be attained in schools across the country that serve disadvantaged populations, this would represent a major success with respect to narrowing the achievement gap, a long-standing national goal that has proven difficult to realize. Number Worlds is not the only curriculum designed to achieve this end. Others include Big Math for Little Kids (Ginsburg and Greenes, 2003) and Children's Math Worlds (Fuson, 2003). While research to compare these curricula on a variety of dimensions is in order, it is clear that the tools to better prepare disadvantaged children for mathematics are now available.

With respect to the second concern—building children's mathematical reasoning ability—controversy persists. While there is evidence that procedural knowledge without conceptual understanding leads to poor mathematical reasoning, it is also well documented that procedural knowledge is a critical element of mathematical competence (National Reasearch Council, 2001a; Haverty, 1999). Without adequate procedural knowledge, not only are children unable to engage in more challenging problem solving, but also, they are unable to engage in basic everyday transactions, like making change. The goal, then, must be one of strengthening mathematical reasoning without sacrificing procedural knowledge.

Research done in the 1990s investigated the effects on student achievement of instruction that builds on informal understandings and emphasizes mathematical concepts and reasoning. Cobb et al.'s problem-centered mathematics project (Wood and Sellers, 1997), and cognitively guided instruction in problem solving and conceptual understanding (Carpenter et al., 1996) both reported positive effects. With support from the National Science Foundation (NSF), several full-scale elementary mathematics curricula with embedded assessments have been developed, directed at supporting deeper conceptual understanding of mathematics concepts and building on children's informal knowledge of mathematics to provide a more flexible foundation for supporting problem solving. Three curricula developed separately take somewhat different approaches to achieving those goals: the Everyday Mathematics curriculum, the Investigations in Number, Data and Space curriculum, and the Math Trailblazers curriculum (Education Development Center, 2001).

All three curricula show positive gains in student achievement in implementation studies, in which the developers collect data on program effects. While such findings are encouraging, they must be viewed with a critical eye, both because those providing the assessment have a vested interest in the outcome and because the methodologies employed generally do not allow for direct attribution of the results to the program.[1] Third party evaluations using comparison groups have been done in some cases, but none of these has involved random assignment, the condition that maximizes confidence in attributing results to the intervention. Nor do these studies measure either fidelity of implementation of the reform curriculum for the experimental group or the specific program features of the alternative used with the control group (see, for example, Fuson et al., 2000).

How students taught with these curricula compare with each other in mathematical proficiency and, perhaps more importantly, how they compare with students taught with curricula that devote more instructional time to strengthening formal procedural knowledge have not been carefully studied. From the perspective of practice, these are important omissions. To make informed curriculum decisions, teachers and school

---

[1]Implementation studies generally do not involve controlled experimentation that allows for comparison of results of one intervention with another. It is also widely understood that the introduction of a new program can have positive effects not because of program content but because something new is being tried.

BOX 3.3 Primary School Mathematics

From an early age, children begin to develop an informal understanding of quantity and number. Careful research conducted by developmental and cognitive psychologists has mapped the progression of children's conceptual understanding of number through the preschool years. Just as healthy children who live in language-rich environments will develop the ability to speak according to a fairly typical trajectory (from single sound utterances to grammatically correct explanations of why a parent should not turn out the light and leave at bedtime), children follow a fairly typical trajectory from differentiating more from less, to possessing the facility to add and subtract accurately with small numbers. But just as a child's environment influences language development, it influences the rate of acquisition of number concepts. For many children whose early years are characterized by disadvantage, there is a substantial lag in the development of the number concepts that are prerequisite to first grade mathematics.

Between the ages of 4 and 6, most children develop what Case and Sandieson (1987) refer to as the "central conceptual structure" for whole number mathematics. The concepts are *central* in the sense that they are vital to successful performance on a broad array of tasks, and their absence constitutes the major barrier to learning. That structure involves four steps (pictured in Figure 3.1) that are developed in sequence:

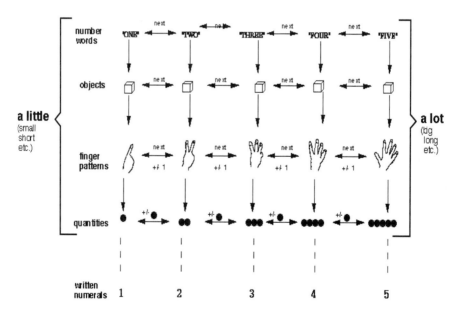

FIGURE 3.1 "Mental counting line" central conceptual structure.

The bottom row of Figure 3.1 indicates that children recognize the written numerals. This information is "grafted on" to the conceptual structure above.

1. **The ability to verbally count using number words.** This ability is initially developed as a sequencing of words (one, two, three . . .) without an understanding of the specific meaning attached to the words. Quantity is still understood nonnumerically as more or less, big or small.
2. **The ability to count with one-to-one correspondence.** When this ability develops, children are able to point at objects as they count, mapping the counting words onto the objects so that each is tagged once and only once. This ability is initially developed as a sensorimotor activity, with an understanding of *quantity* still absent. Children who can successfully count four objects and five objects cannot answer the question, "Which is more, four or five?"
3. **The ability to recognize quantity as set size.** With development of this ability, children do understand that "three" refers to a set with three members. Initially this understanding is concrete, and children will often use their fingers as indicators of set membership.
4. **The ability to "mentally simulate" the sensorimotor counting.** When this ability is in place, children can carry out counting tasks as though they were operating with a mental number line. They understand that movement from one set size to the next involves the addition or subtraction of one unit.

While children with middle and higher socioeconomic status generally come to school with the central conceptual structure in place, many children from disadvantaged backgrounds do not. When first grade math instruction assumes that knowledge, these children are less likely to succeed.

Sharon Griffin and Robbie Case designed a curriculum called Number Worlds that deliberately puts the central conceptual structure for whole number in place in kindergarten (Griffin and Case, 1997). Additional activities extend the knowledge base through second grade. Developed and tested with classroom teachers and children, the program consists primarily of 78 games that provide children with ample opportunity for hands-on, inquiry-based learning. Number is represented in a variety of forms—on dice, with chips, as spaces on a board, as written numerals. An important component of the program is the Number Knowledge Test, which allows teachers to quickly assess each individual student's current level of understanding, and to choose individual or class activities that will solidify fragile knowledge and take students the next step.

The Number Worlds program has been tested with disadvantaged populations in numerous controlled trials in both the United States and Canada with positive results. One longitudinal study charted the progress of three groups of children attending school in an urban community in Massachusetts for three years: from the beginning of kindergarten to the end of second grade. Children in both the Number Worlds treatment group and in the control group were from schools in low-income, high-risk communities where about 79 percent of children were eligible for free or reduced-price lunch. A third normative group was drawn from a magnet school in the urban center that had attracted a large number of majority students. The student body was predominantly middle income, with 37 percent eligible for free or reduced-price lunch.

BOX 3.3 Continued

As Figure 3.2 shows, the normative group began kindergarten with substantially higher scores on the Number Knowledge Test than children in the treatment and the control groups. The gap indicated a developmental lag that exceeded one year, and for many children in the treatment group was closer to two years. By the end of the kindergarten year, however, the Number Worlds children narrowed the gap with the normative group to a small fraction of its initial size. By the end of the second grade, the treatment children actually outperformed the magnet school group. In contrast, the initial gap between the control group children and the normative group did not narrow over time. The control group children did make steady progress over the three years; however, they were never able to catch up.

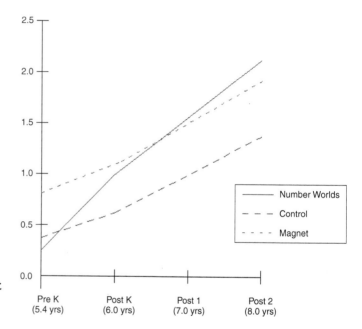

FIGURE 3.2 Mean developmental scores on number knowledge test at four time periods.

administrators need to know what type of implementation of a specific curriculum produces what results, compared with the alternatives before them. Yet to provide the information that is most useful to practice is a major undertaking. These questions are answerable, but research carefully designed to provide those answers will take a substantial investment.

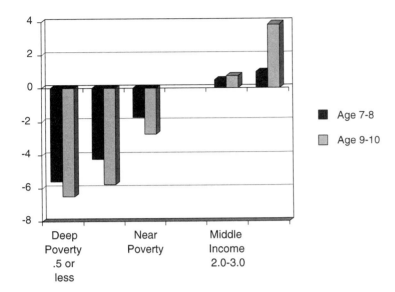

FIGURE 3.3 Income-to-needs and child cognitive ability: Deep poverty and math ability (PIAT-Math), NLSY-CS data set.
SOURCE: Brooks-Gunn et al. (1999).

The significance of these findings is suggested by the data in Figure 3.3 that plots Peabody Individual Achievement Test (PIAT) math scores against poverty level. Clearly the correlation is powerful; the deeper the level of poverty, the poorer the math scores. Importantly, as students move through school, the gap becomes more pronounced. Children ages 9-10 showed even larger score disparities than those ages 7-8. NAEP data indicate that in 1999 black 4- and 8-year-olds ranked in the 15th and 14th percentiles in math, respectively (Thernstrom and Thernstrom, in press). If the Number Worlds program can put poor children on a path to success in math, the contribution would be substantial.

Checkpoints: Assessment

The curricula described above have embedded assessments that allow teachers to track student learning. A key feature of the Number Worlds curriculum is the Number Knowledge Test, which allows teachers to closely link instructional activities for children to the assessment results. How well other curricula link assessment and instruction is an issue worth investigation.

A separate issue is the assessment over time of the five strands that constitute mathematical proficiency. The past decade has seen the emergence of a spate of new tests and measures. No consensus has emerged, however, on critical measures. While there are some standard and widely used assessment tools to appraise young children's emergent reading and language skills and competence, no such tools are used on any comparable basis in primary mathematics.

This type of assessment is required to evaluate the effectiveness of a particular curriculum and to make comparisons across curricula. For the most part, we lack sophisticated methods for tracking student learning over time or for examining the contribution of any particular instructional interventions, whether large or small, on students' learning.

TEACHER KNOWLEDGE

Little is known about what it takes for teachers to use particular instructional approaches effectively, a necessary element of taking any approach to scale. The challenges can be substantial. The curricula mentioned above introduce major changes in approach to teaching mathematics, and effective implementation will require that teachers change their view of mathematics teaching and learning dramatically. In Everyday Mathematics, for example, teachers are expected to introduce topics that will be revisited later in the curriculum. Complete mastery is not expected with the first introduction. This has created some confusion for teachers, who are often unclear about when mastery is sufficient to move on to the next topic (Fuson et al., 2000).

All of the curricula encourage building on students' own strategies for problem solving and supporting engagement through dialogue about the benefits of alternative strategies. The change required on the part of the teacher to relinquish control of *the answer* in favor of a dialogue among students has proven difficult when it has been studied (Palincsar et al., 1989). Adequate opportunities to learn and the ongoing supports for an entirely different approach to teaching will be critical to the effectiveness of efforts to scale up the implementation of the curricula. This is clearly an area in need of further study.

One clue regarding teacher knowledge requirements can be found in research pursued for the most part separately from the work on student learning and the design of curriculum approaches, tools, and materials discussed above. Investigations of teachers' knowledge reveals that although teachers can, for the most part, "do" the mathematics themselves, they often are unable to explain why procedures work, distinguish different interpretations of particular operations, or use a model to closely map the meaning of a concept or a procedure. For example, teachers may be able to use concrete materials to verify that the answer to the subtraction problem in Box 3.1 is 37 and not 43. They can operate in the world of base ten blocks to solve $51 - 14$ but may not be able to use base ten blocks to demonstrate the meaning of each step of the conventional (or other) algorithm.[2]

Similarly, teachers may be able to compute using familiar standard algorithms but not be able to recognize, interpret, or evaluate the mathematical quality of an alternative algorithm. They may not be able to ascertain whether a nonconventional method generalizes or to compare the relative merits and disadvantages of different algorithms (for example, their transparency, efficiency, compactness, or the extent to which they are either error-prone or likely to avert calculational error). These are clearly critical skills if teachers are to work with students' informal understandings and strategies. Over and over, evidence reveals that knowing mathematics for oneself (i.e., to function as a mathematically competent adult) is insufficient knowledge for teaching the subject. In the domain of early number, studies suggest that most teachers' own procedural knowledge is solid, but that their understanding of conceptual foundations is uneven. In a comparative study of elementary mathematics teachers in China and the United States, Liping Ma (1999) found a much larger proportion of U.S. teachers unable to explain whole-number problems—like subtraction with regrouping—using core mathematical concepts (see Box 3.4).

Following this work, some materials for use in teachers' professional instruction have been developed.[3] Modules and other curriculum materials contain focused work aimed at helping teachers learn the sort of mathematical knowledge of whole

---

[2]Base ten blocks are a common material used to model place value concepts, and operations that rely centrally on place value. The materials consist of a unit cube, a ten-stick built of 10 cubes, a flat square built of 100 cubes or 10 ten-sticks, and a block composed of 1,000 cubes, or 10 flats, or 100 ten-sticks.

[3]See, for example, work by Schifter and her colleagues at Education Development Center, Inc., developing mathematics instruction.

BOX 3.4 Organizing Knowledge Around Core Concepts: Subtraction with Regrouping

A study by Liping Ma (1999) compares the knowledge of elementary mathematics of teachers in the United States and in China. She gives the teachers the following scenario:

Look at these questions (52–25; 91–79, etc.). How would you approach these problems if you were teaching second grade? What would you say pupils would need to understand or be able to do before they could start learning subtraction with regrouping?(p.1).

The responses of teachers were wide-ranging, reflecting very different levels of understanding of the core mathematical concepts. Some teachers focused on the need for students to learn the procedure for subtraction with regrouping:

Whereas there is a number like 21–9, they would need to know that you cannot subtract 9 from 1, then in turn you have to borrow a 10 from the tens space, and when you borrow that 1, it equals 10, you cross out the 2 that you had, you turn it into a 10, you now have 11–9, you do that subtraction problem then you have the 1 left and you bring it down. (p.2).

Some teachers in both the United States and China saw the knowledge to be mastered as procedural, although the proportion was considerably higher in the United States. Many teachers in both countries believed students needed a conceptual understanding, but within this group there were considerable differences. Some teachers wanted children to think through what they were doing, while others wanted them to understand core mathematical concepts. The difference can be seen in the two explanations below.

They have to understand what the number 64 means . . . I would show that the number 64, and the number 5 tens and 14 ones, equal the 64. I would try to draw the comparison between that because when you are doing regrouping it is not so much knowing the facts, it is the regrouping part that has to be understood. The regrouping right from the beginning.

This explanation is more conceptual than the first, and it helps students think more deeply about the subtraction problem. But it does not make clear to students the more fundamental concept of the place value system that allows the subtraction problems to be connected to other areas of mathematics. In the place value system, numbers are "composed" of 10s. Students already have been taught to compose 10s as 10 ones and 100s as 10 tens. A Chinese teachers explains as follows:

What is the rate for composing a higher value unit? The answer is simple: 10. Ask students how many ones there are in a 10, or ask them what the rate for composing a higher value unit is, their answers will be the same: 10. However, the effect of the two questions on their learning is not the same. When you remind students that 1 ten equals 10 ones, you tell them the fact that is used in the procedure. And this somehow confines them to the fact. When you require them to think about the rate for composing a higher value unit, you lead them to a theory that explains the fact as well as the procedure. Such an understanding is more powerful than a specific fact. It can be applied to more situations. Once they realize that the rate of composing a higher value unit, 10 is the reason why we decompose a ten into 10 ones, they will apply it to other situations. You don't need to remind them again that 1 hundred equals 10 tens when in the future they learn subtraction with three-digit numbers. They will be able to figure it out on their own (p.11).

Emphasizing core concepts does not imply less of an emphasis on mastery of the procedures or algorithms. Rather, it suggests that the procedural knowledge and skills be organized around the core concepts. Ma describes the set of Chinese teachers who emphasize core concepts as seeing the knowledge in "packages" in which the concepts and skills are related. While the packages differed somewhat from teacher to teacher, the knowledge "pieces" to be included were the same. She illustrates a knowledge package for subtraction with regrouping, which is reproduced below.

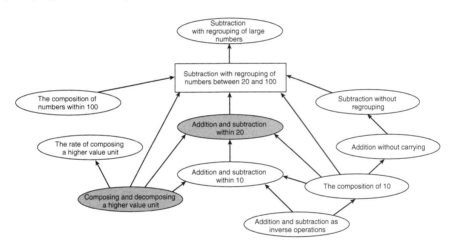

FIGURE 3.4 A knowledge package for subtraction with regrouping.

Two shaded elements in the knowledge package are considered critical. "Addition and subtraction within 20" is seen as the ability that anchors more complex problem solving with larger numbers. That ability is seen as both conceptual and procedural. "Composing and decomposing a higher value unit" is the core concept that ties this set of problems to what the mathematics students have done in the past and to all other areas of mathematics that students will learn in the future.

numbers and operations that is needed for teaching. As with the curricula developed for students' learning discussed above, developers of teacher learning materials do provide some evidence of teachers' learning of mathematics for teaching, but the role of this learning in their instructional practice and effectiveness has not been sufficiently explored. And still less is known about what teacher developers themselves need to know to support teachers' learning and how their professional learning might be supported. The demand for skilled leaders who can teach teachers is growing, and yet the people who play these roles are more varied than any other category of educators and often have no professional preparation for working with teachers. Scaling up materials that can support teachers' learning of mathematics for teaching will require attention to the knowledge requirements of those who will guide and support teachers.

RESEARCH AGENDA

Given the current state of practice and knowledge about learning and teaching of early number, then, what might a SERP program of research and development seek to do? How might it build on what currently exists and begin to extend and fill gaps in what is known and done, with the ultimate goal of more reliably and productively building evidence-based instructional practice? In other words, how could work be planned and carried out that would extend what is known and take that to scale in U.S. schools?

The proposed agenda is comprised of three major initiatives:

1. focus on developing assessments to measure student knowledge;
2. evaluation of promising curricula and the effects of their particular design features on student outcomes;
3. focus on the teacher knowledge requirements to comfortably and effectively use curricula that are built on research-based findings regarding student learning.

Initiative 1: Developing Early Mathematics Assessments

As described in Appendix A, quality assessments depend on three things: (1) clarity about the competencies that the assessment should measure, (2) tasks and observations that effectively capture those competencies, and (3) appropriate qualitative and quantitative techniques to give interpretive power to the test results. Clarity about the competencies to be measured requires a theoretical understanding (that is empirically supported) of mathematics learning. Unlike many other areas of the curriculum, early mathematics has the theoretical and conceptual models, as well as supporting empirical data, on which to build quality assessments. Substantial work has already been done to specify critical concepts and skills in this domain, providing assessment developers with substantial resources on which to draw in drafting the elements of a measurement strategy.

Even with a strong foundation on which to build in early mathematics, much work remains in designing and testing assessment items to ensure that inferences can accurately be drawn about student knowledge and competencies. And this work must be carefully crafted for the specific purpose and use of the assessment. This includes formative assessment for use in the classroom to assist learning. These assessments can, for example, provide feedback to the teacher on whether a particular skill or concept is mastered adequately, or whether some individual students need more time and practice before moving on. Summative assessments are also used in the classroom, but they come at the end of a unit, giving a teacher feedback on how well the students have mastered and brought together the set of concepts and skills taught in the unit. These may be helpful to the teacher in redesigning instruction for the next year, providing valuable data on students' strengths and weakness that can inform instruction at the next level. School- or district-level assessments have a separate purpose. Rather than provide information on individual students, their purpose is to determine attainment levels for students as a group in order to evaluate the effectiveness of the school's or the district's instructional program and, in some cases, to hold schools accountable for the performance of their students.

Currently the different types of assessment are loosely connected at best. Tensions are introduced when strong instructional programs and accountability assessments are at odds. Better aligning assessments, tying all assessments firmly to the theoretical and empirical knowledge base, are widely regarded as important to improving learning outcomes. In the area of mathematics, SERP affords a unique opportunity to pursue the development of an integrated assessment system with the three critical characteristics of comprehensiveness, coherence, and continuity described in Chapter 1 (National Research Council, 2001c). The construction of such a system is a major research, development, and implementation agenda that would require the stability, longevity, and support that SERP intends as its hallmark.

The work should be pursued as a collaborative effort involving teachers, content area specialists, cognitive scientists, and psychometricians. The effort could use as a departure point well-established standards in mathematics (e.g., National Council of Teachers of Mathematics), standards-based curricular resources, and rigorous research on content learning to identify and define what students should know in early mathematics, how they might be expected to show what they know, and how to appropriately interpret student performance. In the case of formative assessment, this extends to an understanding of the implications of what the evidence suggests for subsequent instruction. In the case of summative assessment, this means understanding the implications of student performance for mastery of core concepts and principles and its growth over time.

While there are several possible approaches to developing such a system of student assessments in early mathematics, one obvious place to begin is with a review of the assessment materials in existing widely used and exemplary curricular programs for formative and summative assessments, as well as state and national tests for policy making and accountability. These can be reviewed in light of cognitive theories of mathematical understanding, including empirical data regarding the validity of specific assessments. Research needs to focus on evidence of the effectiveness of specific assessments for capturing the range of student knowledge and proficiency for particular mathematical constructs and operations. A related line of inquiry should focus on issues of assessment scoring and reliability, particularly ease of scoring, consistency of scoring within and across individuals, and consistency of interpretation of the results relative to the underlying cognitive constructs.

The power offered by assessments to enhance learning depends on changes in the relationship between teacher and student, the types of lessons teachers use, the pace and structure of instruction, and many other factors. To take advantage of the new tools, many teachers will have to change their conception of their role in the classroom. They will have to shift toward placing much greater emphasis on exploring students' understanding with the new tools and then applying a well-informed understanding of what has been revealed by use of the tools. This means teachers must be prepared to use feedback from classroom and external assessments to guide their students' learning more effectively by modifying the classroom and its activities. In the process, teachers must guide their students to be more actively engaged in monitoring and managing their own learning—to assume the role of student as self-directed learner. Clearly research of this type focuses on issues of teacher learning and knowledge that are poorly understood at present.

The development of assessments in early mathematics should therefore be closely tied to complementary initiatives in the areas of teacher knowledge and curriculum effectiveness. A strand of research focused on implementation issues should address the set of questions critical to successful use of quality assessments:

- What teacher knowledge is necessary to support effective use of assessments in their instructional practice? These include teacher understanding of the assessments and their purpose, as well as practical considerations of the time to administer, score, and interpret results;
- What forms of technology support are needed to assist teachers in the administration, scoring, and interpretation of a range of standards-based and theory-based assessments; and
- How, and to what extent, does the process of implementing curriculum-based and standards-based assessments lead to changes in teachers' instructional practices, and how do these changes affect student learning outcomes? This investigation should focus both on

changes in the near term and the stability of changes in the long term.

The power of new assessments to support learning also depends on substantial changes in the broader educational context in which assessments are conducted. For assessment to serve the goals of learning, there must be alignment among curriculum, instruction, and assessment. Furthermore, the existing structure and organization of schools may not easily accommodate the type of instruction users of the new assessments will need to employ. For instance, if teachers are to gather more assessment information during the course of instruction, they will need time to assimilate that information, thus requiring more time for reflection and planning. As new assessments are implemented, researchers will need to examine the effects of such factors as class size and the length of the school day on the power of assessments to inform teachers and administrators about student learning. It will be important for the work on learning and instruction to be closely tied to the work on schools as organizations.

Initiative 2: Teacher Knowledge

To take advantage of existing investments in research and development in elementary mathematics will require further work regarding teacher learning and knowledge requirements. There are two approaches to this teacher learning that could strategically build on work that has already been done. The first emphasizes teachers' understanding of mathematical concepts and the connections among them, and the second focuses on the knowledge required for teachers to use promising curricula comfortably and effectively.

The work of Liping Ma described earlier (see Box 3.4) suggests that Chinese teachers who have a strong conceptual understanding of elementary mathematics organize their knowledge into "knowledge packages" in which core concepts are central and other required understandings and skills are organized around these concepts. Ma's study is qualitative and so does not offer empirical support for the notion that the conceptual understanding of Chinese teachers is what produces higher student achievement. The power of the study is supported, however, by its theoretical underpinnings. Research on human learning suggests that the organization of knowledge around core concepts is a key component to the effective acquisition, use, and transferability of that knowledge (National Research Council, 2000).

Ma's work provides a point of departure for research that further explores the notion of knowledge packages and incorporates them into the preparation of elementary mathematics teachers. That research should be designed to measure the effects of that instruction at its conclusion, as well as at several periods of time after a teacher has engaged in practice. The research should be done in carefully controlled trials to allow for attribution of the outcomes to the program.

The second exploration of teacher knowledge should build on research that suggests that professional development is more productive when it is tied to specific curricula or instructional programs that teachers will then incorporate into their practice (Cohen and Hill, 2000). This research should begin with a clear articulation of the principles and assumptions about student learning that the curriculum incorporates, and comparing these to carefully solicited understandings of teachers. Learning experiences should be designed to address the points of divergence and tested for their power to change teacher conceptions.

In order to support taking curricula to scale, the support for teachers must be adequate when a researcher is not involved. This work on teacher knowledge should experiment with levels of support for teachers and involve cycles of design, testing, and redesign to create the materials and other supports that can allow teachers to use the curricula effectively independently.

The research on teacher learning should test effectiveness with regard to *both* teacher learning and the learning of their students. The relative benefits of teacher guides, videotaped cases, and opportunities to pose questions and receive support should be tested, as well as the timing effect (before instruction begins, during instruction, etc.) for different teacher learning opportunities.

Initiative 3: Curriculum Evaluation

The identification (and further development) of a set of approaches to the teaching of number and operations that vary on distinct and theoretically important dimensions would permit careful comparisons of how particular instructional regimes impact students' learning. Programs and approaches designed to build on students' informal mathematical reasoning abilities, such as Number Worlds, Cognitively Guided Instruction, and the three NSF-supported curricula mentioned above, should be compared with more traditional curricula, like those produced by Saxon Publishers, Harcourt Brace, and McDougal Littell. This initial core would be expanded over time to include other theoretically and practically important alternatives.

Many of the evaluations of the curricula set out to answer the question "Does the curriculum improve student achievement?" While this is an important question for schools choosing a curriculum—and of particular interest to those who market a curriculum—the questions of importance for long-term improvements in practice are why, for whom, and compared with what? Exploring why an approach works can provide teachers and curriculum developers with critical information for improving their work. This kind of research will require going beyond evaluation of a curriculum as a whole into experimentation with particular features.

For example, Aleven and Koedinger (2002) compared groups of students who engaged in mathematics problem solving for the same amount of time, but one group was instructed to do self-explanation for each problem, and the other was not. Because self-explanation takes time, the latter group practiced almost twice as many problems, but their learning was more shallow and did not transfer as well as the self-explainers. Understanding the individual contributors (like self-explanation) to outcomes will require this level of probing.

School decision makers also need a knowledge base that will allow them to make more informed choices for their particular school(s). Number Worlds shows very promising results for disadvantaged children; Everyday Mathematics does as well. How, and for whom, do those outcomes differ? Are there trade-offs in the competencies children gain from each? Does the context in which they work best differ? Each of the three NSF elementary mathematics curricula takes a somewhat different approach to instruction. How are those differences reflected in outcomes for students? Does one better address the needs of low- or high-achieving students?

An analysis of existing candidate materials could illuminate important differences. The implementation, adaptation, and use of these different approaches could be followed over time, attending to instructional practice, students' opportunities to learn, and implementation issues. In addition, based on what is known about teachers' knowledge of whole numbers and operations for teaching, as well as about their learning, systematic variations could be designed to support the implementation of these different instructional approaches. For example, in one set of schools, a teacher specialist model might be deployed and, in others, teachers might engage in closely focused study of practice (instruction, student learning, mathematical tasks), coplanning and analyzing lessons across the year. In still others, teachers might be given time and be provided incentives to spend time planning with the ample teacher guides.

The work could be conducted in carefully controlled, longitudinal studies carried out in SERP field sites. Because SERP would have relationships established with a number of field sites and data collection efforts in those sites already under way, taking on a controlled experimental study of alternative curricula would be far less daunting a task than it would be for researchers working independently. Moreover, the concern for undertaking research that is maximally useful to educational practice and the ability to design and conduct—or oversee the conduct of—that research will be combined in a single organization. Such a situation does not now exist.

· ·

ALGEBRA

STUDENT KNOWLEDGE

Algebra, foundational to so much other mathematics, and so poorly learned in general, is an area in critical need of concentrated research and development. Algebra is crucial to the development of mathematical proficiency because it functions as the language system for ideas about quantity and space. Algebra moves attention from particular numerical relations and computational operations to a more general mathematical environment with notation and representation useful across all areas of mathematics. These represent vital, but precarious, passages; students' transitions into the domain of algebra are often plagued with problems.

Although it has long been considered important, attention

to the role of algebra has never been more intense. Traditionally, high school mathematics tended to have two tracks. The academic track included a Euclidean proof-oriented geometry course and a series of algebra courses focused on preparing students for calculus. The other was a nonacademic series of practical courses—sometimes called business math—which often were just a review of middle school mathematics, particularly rational number arithmetic. A series of studies found a strong positive correlation between participation in the academic mathematics stream and future earnings (Pelavin and Kane, 1990). Some studies have shown that this correlation may be causal (Bednarz et al., 1996). Children who would have been directed to the practical stream were successful in the academic stream (Porter, 1998).

All current curricular recommendations and frameworks place a high premium on algebra. The recent National Research Council report, *Adding It Up*, recommends that the basic ideas of algebra as generalized arithmetic be introduced in the early elementary grades and learned by the end of middle school. This is consistent with curricular opportunities in other countries (National Research Council, 2001a).

The Destination: What Do We Want Students to Know and Be Able to Do?

Algebra provides powerful abstract concepts and notation to express mathematical ideas and relationships and a set of rules for manipulating them. These tools are invaluable for solving a wide range of problems. Learning to make sense of and operate meaningfully and effectively with these tools is a central goal of instruction. This power involves both moving from contexts to abstract models and, conversely, interpreting abstract ideas skillfully in concrete situations. Moving beyond this core goal, however, one moves into contested terrain.

What should be the subject matter of algebra? Many new curricula take a view of algebra that is at variance with how it is conceived within the discipline and with how it has been treated historically in the school curriculum. In recent years, a number of individuals, groups, committees, and task forces have worked to provide definitions or characterizations to describe the continually evolving school algebra. These characterizations typically include such ideas as: algebra as a consolidation of, or generalization of, ideas in arithmetic; algebra as the study of structures, patterns, and symbolic representations; algebra as the study of functions, covariation, and modeling.

Overlaid on the content debates are two different visions of how algebra should be taught. The first is motivated by the observation that students do not seem to really understand or value what they are doing in algebra. It therefore emphasizes algebra as a tool for solving real-world problems. In a somewhat caricatured form it might be seen as teaching students to be intelligent spreadsheet users who can create effective formulas and graphs. The second, more traditional view stresses algebra as a preparation for traditional college mathematics (which itself is in a state of reform and counterreform). In its extreme it emphasizes formal proofs and skills that are undoubtedly challenging for most high school students.

Algebra, of course, legitimately includes all of the content areas described above. Both the ability to solve real-world problems and to solve formal proofs are valuable outcomes for different purposes. Trade-offs become necessary only when the limits on instructional time force them. Currently there is little understanding of the affordances of different instructional emphases. It is therefore of little surprise that there is no consensus on how choices should be made or whether different options should be available for pursuing academic algebra.

The Route: Progression of Student Understanding in Algebra

Traditionally, students take a prealgebra course in the eighth grade and go on to a full algebra 1 course in the ninth grade, geometry in the tenth grade, Algebra 2 in the eleventh grade, and elementary functions in the twelfth. However, there are many variations on this pattern, including a substantial minority of students who begin this sequence a year early and many school systems that teach an integrated curriculum from grades 9 to 11 and do not have a year just for geometry. Traditionally, a large fraction of students never even begin this academic stream, and many who do fail to complete it.

Mastery of algebra builds on mastery of the mathematics taught in earlier years. One might wish that students coming into this curriculum would have mastered rational numbers, but this is often not the case and it causes difficulties for mastery of algebra. For instance, it is common to see students perform flawlessly when solving simple linear equations involving integers but fail when the same equations involve fractions. Poor fluency with rational number arithmetic and arithmetic more generally adds to the cognitive demands on a student and interferes with learning of higher level ideas (Haverty, 1999). And poor understanding of rational numbers spells disaster when students come to later topics like rational expressions.

Students also often come with an understanding of their earlier curriculum that creates difficulty for algebra. For instance, the equal sign has typically had the meaning of an operation (e.g., 3 + 4 = is a request to perform addition) rather than of a relationship between two expressions. Still, students have informal ways of reasoning about problems that can be quite powerful. For instance, consider the following two problems:

As a waiter, Ted gets $6 per hour. One night he made $66 in tips and earned a total of $81.90. How many hours did Ted work?

Versus

$x * 6 + 66 = 81.90$

While only 42 percent of algebra 1 students can solve the second equation, a full 70 percent can solve the word problem (Koedinger and Nathan, in press). Students bring informal ways of reasoning about their mathematical knowledge that enables them to solve these problems. Often students have difficulties understanding why they are being taught algebraic symbol manipulation to solve such problems when initially this just makes solving them harder and more error-prone. Rather than replacing the informal with the formal, Koedinger and Anderson (1998) and Gluck (1999) have shown that by building on these informal processes, one can more effectively teach students to use the formalisms of algebra.

How do students develop the capacities to move from contexts to abstract models and, conversely, to interpreting abstract ideas skillfully in concrete situations? Beliefs abound about the directionality of these connections in learning: Some argue that all meaningful learning must move from the concrete to the abstract; others insist that the power of the generalized, abstract forms affords learners greater insight. Notions of what counts as concrete or abstract remain vaguely and variously defined.

The Vehicle: Curriculum and Pedagogy

New curricular materials introduce algebra in various segments of the K-12 mathematics sequence using a variety of conceptions of algebra. Some take an equation-solving orientation; others take a function approach; still others take a mathematical modeling approach. Yet we do not know how these various curricular approaches affect students' understanding and continued use of algebra.

The hypothesis that algebra instruction moves to abstraction without first connecting the abstractions with the concrete instances that justify them has led to the development of new curricula that emphasize richly contextualized problems. Some research evidence suggests that student engagement is higher and that students work meaningfully with important mathematical ideas, outperforming students whose curricular experiences do not include such rich investigative problems (e.g., Boaler, 1997; Brenner et. al., 1997; Nathan et. al., 2002). But some caution that such problems, if taken seriously, demand close attention to the contexts; whereupon students may become preoccupied with the contextual particularities in ways that distract from the mathematical ideas entailed (Lubinski et al., 1998). Consequently, they may not develop abstract knowledge central to mathematical proficiency. Some instructional approaches look for a middle ground in which algebra knowledge is contextualized, but the context is kept simple, and a single context is used extensively to help students see through to the underlying mathematical functions (Kalchman et al., 2001; Kalchman and Koedinger, forthcoming). Much remains to be investigated about how students develop the ability to work effectively with abstract ideas and notation, as well as about the relationships between abstraction and concrete experiences in learning.

A focus on algebra would afford opportunities to probe how different instructional uses of technology interact with the development of symbol manipulation skills. With the increased availability of technology, what is meant by "symbolic fluency" raises new questions. What is the role that graphing calculators and computational algebraic systems might play? What is the role of paper and pencil computation in developing understanding as well as skill? These are questions that appear at every level of school mathematics.

Checkpoints: Assessment

Algebra represents a major challenge for many students. If more students are to succeed in meeting that challenge, it will be important to identify the points of difficulty for individual students and provide effective instructional responses before they are lost. The difficulty factors assessments of algebra reading (Koedinger and Nathan, In Press) and algebra writing (Heffernan and Koedinger, 1997, 1998) are examples of efforts to provide assessment tools for this purpose.

Two features of the subject make assessing individual progress very important. Algebra requires facility with much of the mathematics that has come before. If the mathematical foundation is weak in any of its components, algebra mastery will be undermined. Determining where students need to shore up the preparatory mathematics, as well as opportunities for doing so, are critical to success.

Second, algebra instruction moves toward high-level abstraction. The readiness of individual students to move from one level of difficulty to the next will differ. If the movement comes before a bridge is effectively built to a student's prior knowledge or before new knowledge is consolidated, the student will be lost. If movement toward greater difficulty does not come soon enough, a student will make less progress in higher level algebra than is possible. Indeed, precisely this is at the heart of opposing views of algebra pedagogy. If formative assessment were sophisticated enough to determine individual student readiness to move on, then trade-offs between attending to the needs and preparedness of different students would not be necessary.

A research and development effort at Carnegie Melon University that generated the Algebra Cognitive Tutor has focused very productively on the second element of this problem (see Box 3.5). It began as a project to see whether a computational theory of thought, called ACT (Anderson, 1983), could be used as a basis for delivering computer-based instruction. The cognitive theory applies to problem solving more broadly. For purposes of algebra teaching, it was the foundation for modeling the variety of different approaches—both correct and incorrect—that students typically take to solving algebra problems. A number of different approaches can lead to a correct solution, and the program does not favor one over another. However, some approaches lead the student astray. If the student is working effectively on a problem, there is no computer feedback. But when a student begins down an unproductive or erroneous path, the computer program recognizes this by a process called model tracing and provides hints and instruction to guide the student's thinking.

The Algebra Cognitive Tutor also assesses mastery of elements of the curriculum by a process called knowledge tracing. When a student's problem-solving efforts suggest that the relevant knowledge or skill is not yet consolidated, the computer selects instruction and problems appropriate to where that student is in the learning trajectory.

In studies of cognitive tutors more generally, it was found that under controlled conditions, students could complete the curriculum in about a third of the time generally required to master the same material, with about a standard deviation (approximately a letter grade) improvement in achievement (Anderson et al., 1995). In real classrooms, the impact has generally not been as large. A third-party evaluation of the tutors suggested that the scaffolding of learning that allowed students to experience success with challenging problems produced large motivational gains (Schofield et al., 1990).

TEACHER KNOWLEDGE

In the past, only teachers of high school students were thought to need knowledge of algebra. Although their preparation to teach does not include study of the objects and processes of high school algebra, little attention has been paid to whether or not secondary school teachers do in fact have adequate algebraic knowledge for teaching (Ferrini-Mundy and Burrill, 2002). Students preparing to be secondary school teachers typically take courses in abstract algebra and analysis, under the assumption that such mathematical background will serve them well as secondary school teachers. Yet the actual knowledge developed in such courses and its application by teachers in classrooms has not been thoroughly studied. Some research suggests that the

BOX 3.5 The Algebra Cognitive Tutor

The Algebra Cognitive Tutor is one of a set of cognitive tutors developed at Carnegie Mellon. Of great relevance to the SERP vision, the tutors are a good illustration of how to make the transition from the laboratory to the classroom. The work at Carnegie Mellon began as a project to see whether a computational theory of thought, called ACT (Anderson, 1983), could be used as a basis for delivering computer-based instruction in algebra. The ACT theory of problem-solving cognition is the basis for modeling students' algebra knowledge. These models can be captured in a computer program that can generate or identify a range of characteristic approaches to solving an algebra problem. These cognitive models enable two sorts of instructional responses that are individualized to students:

1. By a process called model tracing, the program will infer how a student is going about problem solving and generate help and instruction appropriate to where that student is in the problem.
2. By a process called knowledge tracing, the program will infer where a student falls in the learning trajectory and select instruction and problems appropriately.

Developing cognitive models that accurately reflect competence and developing appropriate instructional responses is an iterative process. The success of the tutors depends on a design-test-redesign effort in which models are assessed for how well they capture competence and instructional responses are assessed for how effective they are.

In studies of cognitive tutors more generally, it was found that in controlled laboratory condition students using a cognitive tutor could go through a curriculum in a third of the time, and in carefully managed classrooms students would show about a standard deviation (approximately a letter grade) improvement in achievement compared with students receiving standard instruction (Anderson et al., 1995). In real classroom situations, the impact of the tutors tends not to be as large, varying from 0 to 1 standard deviation across more than 13 evaluations. Another third-party evaluation, focusing on the social consequences of the tutors, documented large motivational gains resulting from the active engagement of students and the successful experiences on challenging problems (Schofield et al., 1990).

However, unlike many such small-scale success stories in cognitive science, this project was able to grow to the point at which the cognitive tutors now are used in over 1,200 schools, 46 of 100 largest school districts, and interacting with about 170,000 students yearly. A number of features were critical to making this successful transition:

1. While the ACT theory provided the technology, there was a concerted effort to identify a curriculum that educators wanted to be taught in the classroom. In particular, the project recognized that it was a priority for the schools to teach a curriculum that was in compliance with the NCTM standards (National Council of Teachers of Mathematics, 2000) and designed a curriculum around this.

2. A curriculum was designed that teachers would accept and could implement. A full-year curriculum was developed rather than an enrichment program to be inserted into an existing curriculum. The curriculum was designed with the critical help of teachers with experience in urban classrooms. The computer tutors were used as a support rather than a replacement for the teachers. In this curriculum students spend 40 percent of their time with the computer tutors and 60 percent of their time with other activities. These classroom activities help them transition to their lessons with the tutor and transition those lessons to mathematics that they will have to do without the tutor on paper and in the real world.

3. A structure was set up for supporting the use of the curriculum and tutors. Before introducing the tutors into a classroom, it has been important to provide professional development time to enable teachers to prepare for the change they are about to experience. A center at Carnegie Mellon was set up for responding to teacher and school problems. As the adoptions grew, a separate company, Carnegie Learning, was created to perform this function and maintain and adapt the materials.

4. Ultimately, such a curriculum must be financially self-sustaining and it was developed from the beginning with a plausible financial model in mind. In particular, by offering a full grade 9-11 curriculum, it was possible to earn in sales the kind of income that is necessary to sustain this activity.

While the cognitive tutor enterprise illustrates what needs to be done to transition research ideas into the American classroom, it does not represent a complete solution to even high school algebra. Early in the development of the Algebra Cognitive Tutor, a decision was made to place a heavy emphasis on contextualizing algebra to help students make the transition to the formalism. The course has been demonstrated to raise student achievement in urban schools and to reduce the number of students dropping out. However, high-achieving students may not be fully achieving the desired fluency in symbol manipulation and abstract analysis. There is no reason why the cognitive tutors could not more fully address these topics and, in fact, many tutor units do, particularly in the algebra 2 course. However, a more accelerated course may yield better results for high-achieving students.

assumption that secondary school teachers have strong and flexible knowledge of algebra is unfounded (Ball, 1988). In fact, evidence suggests that secondary school teachers may often have rule-bound knowledge of procedures but lack conceptual, connected understanding of the domain. That some signals existed that high school teachers' knowledge may not be as robust as had been suspected is not so surprising, since they are educated in the very mathematics classrooms that many seek to improve.

Still, this result signals a more significant problem. First, the changes in and expansion of what is meant by algebra mean that secondary school teachers are increasingly being called on to teach algebraic ideas and connections that they have not themselves studied or have not studied in such ways. Analyses of what the new curricula demand of teachers could make visible the mathematical demands of those curricula and permit investigation of teachers' current knowledge to teach those materials. Second, the movement of algebraic ideas into the middle school and especially the elementary school curriculum means that teachers who have not in the past taught algebra are now being called on to teach ideas and processes for which they have not in the past been responsible. Prospective elementary school teachers' knowledge of algebra may be based largely on their own experiences as high school students. The new requirements of elementary school teaching raise important and pressing needs for research on teacher knowledge, teaching, and teacher learning specifically in algebra.

Not only do teachers need knowledge of the mathematical content, however. Equally important (and related) is knowledge of how students think about and develop algebraic ideas and processes. What ideas or procedures are particularly difficult, both in reading and writing mathematical relationships, for many students? As algebra shifts to being a K-12 subject, rather than a pair of high school courses, new questions emerge that warrant investigation: If students learn about variables and equations sooner and engage earlier in algebraic reasoning (Carpenter and Franke, 2001), how will these earlier experiences shape the development of students' algebraic proficiency over time? How do students of different ages manage and use symbolic notation, both in reading and writing mathematical relationships? What supports the development of meaningful and skilled fluency with mathematical symbols and syntax? In facing diverse classes of students, teachers also need to understand better the mathematical resources and difficulties that their students bring from their own environments, as well as how to make productive use of and mediate those (see Moses and Cobb, 2001, for a robust example of designing strong connections between the domain of algebra and students' out-of-school activities and knowledge use).

RESEARCH AGENDA

There is little agreement at this point on what algebra should be taught or how it should be taught. As in other areas of the curriculum, the questions are in part a matter of valued outcomes for algebra instruction and the instructional time allocation across algebra and other subjects. But a study of the outcomes of different instructional choices can make the decisions far more rational than they can be in the absence of high-quality data.

We propose research and development on four major initiatives in this area:

1. Alternatives in the teaching and learning of algebra;
2. Teacher knowledge;
3. Developing algebra assessments and instruments;
4. Students' development over time and the effects of different curricular choices.

Initiative 1: Research and Development on Alternatives in the Teaching and Learning of Algebra

Work supported by the National Science Foundation as well as by private foundations has generated a variety of curriculum materials for schools that constitute different perspectives on algebra, different ideas about what is important for students to learn, and different ideas about how students can most effectively be taught that can be contrasted with the best traditional approaches to teaching algebra. Since these curricula are already developed and in use, they provide an opportunity for understanding the consequences of the choices made.

For example, in some materials a functions approach to algebra is central, while in others, algebra is treated more as generalized arithmetic, and the solving of equations is more prominent. In some approaches, students are engaged in using the tools of algebra to model situations and problems, while, in others, algebra as an abstract language is stressed. While much controversy surrounds the worth and merit of these different perspectives on the subject, additional debates center on the contribution of calculators and other technology, the structure of lessons, and the role of the teacher. Because curricula have already been developed that represent these different perspectives on the subject and on how it might best be taught, one important initiative of SERP might be to design comparative studies of how these curricula are taught in classrooms and what and how diverse students learn algebra over time.

In this initiative, cohorts of students could be followed longitudinally. Studies could gather information about the instruction they receive, exposure to curriculum, information on the teachers, and their use of the curriculum and other tools. This initiative will depend on the development of effective assessments (see Initiative 3).

As with elementary mathematics, however, knowing why particular curricular interventions produce particular outcomes will require companion controlled experiments at the level of particular program features to test for causality. This kind of research is necessary not only to advance scientific understanding, but also because it provides critical knowledge for teachers who adapt curricula and allows developers to improve curricula or design alternatives that are responsive to research findings.

Simultaneous with this effort, SERP can support curriculum development that extends existing curricula in promising directions. The Algebra Cognitive Tutor, for example, emphasizes highly contextualized problem solving. While many fewer students drop out and students master the material covered more quickly and effectively, the curriculum may not achieve the fluency in symbol manipulation and abstract analysis expected for high-achieving students. The developers suggest that the curriculum could quite easily be strengthened in this respect, and a separate accelerated algebra course is likely to yield even better results for high-achieving students. In studying the set of curricula as they are being implemented, SERP as a third-party entity would be well positioned to identify and support promising areas like this for further development.

Initiative 2: Research and Development on Teacher Knowledge

Important questions remain unanswered about the knowledge of mathematics needed to teach algebra effectively. As with elementary mathematics, the existence of different curricular approaches and efforts to study them, as outlined above in Initiative 1, provide the opportunity to investigate the demands for teachers in teaching different curricular approaches to algebra. For example, specifically what mathematical demands arise for teachers in teaching approaches to algebra that emphasize symbolic fluency compared with approaches that emphasize modeling and connections to situations? What sort of representational and notational fluency do teachers need? How do teachers need to understand the connections between algebra and other domains of mathematics, and what is demanded of teachers with respect to mathematical reasoning under different approaches to algebra?

The movement of algebra into the elementary school curriculum, as recommended both by the National Council of Teachers of Mathematics' *Principles and Standards for School Mathematics* (2000) and *Adding It Up* (National Research Council, 2001a), creates the opportunity to examine what elementary teachers need to know with respect to algebra. Typically regarded as a secondary school subject, algebra has not played a central role in the preparation of elementary school teachers. Studies of teachers engaged with the new curricula that include elementary school skills and ideas of algebra could provide insight into the kinds of algebra knowledge useful to the teaching of young children. Where and how do ideas and skills of algebra surface in younger children's learning, and what sorts of knowledge would help teachers address and develop those? As in other areas of the curriculum, it will be particularly important to identify the issues with which teachers struggle most, the conceptions that make effective teaching more difficult.

As in Initiative 1, the study of teacher knowledge requirements would provide the basis for research and development on effective teacher education interventions. The development efforts would be expected to target a variety of teacher learning opportunities, including pre-service education in teaching mathematics, teacher support materials, and in-service education associated with the use of particular curricula.

Initiative 3: Developing Algebra Assessments and Instruments

Efforts to improve algebra instruction, as well as to evaluate the effectiveness of alternative approaches to the teaching of algebra, will depend on the development of new assessments of students' and teachers' learning.

The development of formative assessments for instructional purposes will need to test hypotheses about what is difficult for students to learn, as well as hypotheses about the kinds of scaffolds that provide support for learning when students are struggling. For classroom effectiveness, these assessments must be closely tied to instructional materials. An investment in the development of algebra assessments that capture all aspects of algebra proficiency, including the robustness and flexibility of conceptual and procedural knowledge and the ability to transfer learning to novel problems, will need to be developed if outcomes of alternative approaches to instruction are to be meaningfully compared.

Assessments will also be needed that can discriminate different kinds and levels of knowledge for the teaching of algebra. These should include both the knowledge of subject matter and pedagogical content knowledge.

Moreover, in order to compare differences in students' opportunities to learn in circumstances in which the teacher or the curriculum changes, instruments to gather information about instruction itself will be important. For example, the representations and tools that are used and the type, frequency, and duration of their use needs to be captured. Measures of fidelity of implementation and of teacher support will be required as well. These investments in instrumentation and assessment tools at the start will allow for subsequent work to be far more powerful for guiding instructional practice.

Initiative 4: Students' Development Over Time and the Effect of Different Curricular Choices

Because algebra is increasingly seen as a K-12 strand of a mathematics curriculum, not merely as a high school course or pair of courses, the timing is right to design studies that track students across their school careers, investigating the development of proficiency in algebra. Such longitudinal studies of algebra learning could be designed to examine how particular configurations of curricular and pedagogical choices affect what students learn. For example, do students whose experiences with number and operations are designed to develop deep conceptual understanding and procedural fluency fare differently in algebra than those whose opportunities to learn emphasize applications and modeling? How do differences in the development of arithmetic fluency affect the development of students' algebraic proficiency?

Initially, the work involved will be to design careful procedures for longitudinal data collection. Doing so will hinge on Initiative 3, in which input and outcomes measures are tested and developed. While the fruits of this research would not be expected in the early years of the program, designing the data collection effort early and carefully will be critical to high-quality analysis further down the road.

4 Science

Many young children get off to a good start in acquiring knowledge on a variety of scientific topics. U.S. fourth graders score near the top in science—just behind Korea and Japan—among the nations in the Third International Mathematics and Science Study (TIMSS) (National Center for Education Statistics, 1999). However, since little science is taught in the early elementary grades, it is unlikely that these results can be attributed to school science programs. Instead, the high scores probably reflect the many informal science learning opportunities that abound in the United States, including science and technology museums, youth organizations that support science activities, television (e.g., the Discovery Channel; Magic School Bus; 3-2-1 Contact; Bill Nye, the Science Guy), trade books, and children's science magazines.

When serious science instruction begins, typically in middle school or even later, the advantages of informal learning resources begin to be overtaken by the disadvantages of unfocused curricula and weak teacher knowledge of both science content and pedagogy. At this stage the international comparisons become much less favorable. In fact, the TIMSS results at grade 8 place the United States in 17th place out of 26 nations. By grade 12 the United States scores are lower still, with advanced U.S. students scoring last of the 16 countries compared. Scores on national assessments confirm the bleak TIMSS results. On the 2000 National Assessment of Educational Progress (NAEP), 47 percent of twelfth grade students scored in the lowest category (below basic proficiency), an increase from 43 percent in 1996 (National Center for Education Statistics, 2003). Clearly, science education is not on a path to improvement.

As in reading and mathematics, there are pockets of research and development in science education that have produced instructional programs with demonstrated student achievement benefits. In physics, for example, a highly productive tradition of research has produced a deep knowledge base with very important implications for educational practice. In contrast, for other areas in science education the research base is not yet developed fully enough to guide or support decisions about instruction. As with reading comprehension, knowledge of the progression of student understanding is relatively sparse and spotty from topic to topic. Moreover, there is very little evidence about how student understanding can develop with instruction over the school years.

The first section of this chapter, as in the chapters on mathematics and reading, addresses an area that has potential for wide impact in the relatively near term: physics. Unlike the other two disciplines, however, this downstream case falls late in the K-12 curriculum. The second section of this chapter addresses science education across the school years, since we are still far upstream in developing a principled organization of science instruction, particularly in the years before high school.

THE TEACHING AND LEARNING OF PHYSICS

The number of students who take courses in physics is relatively small in comparison to the number who take biology or chemistry, as is the number of credentialed high school physics teachers in comparison to other science teachers. Perhaps because the community of educators working on physics is small, it has been possible to pursue a cumulative research agenda on major issues in physics teaching and learning.

STUDENT KNOWLEDGE

The Destination: What Should Students Know and Be Able to Do?

Until relatively recently there was substantial overall agreement regarding what students should know and be able to do in the typical high school or college physics course (the content of the two overlaps substantially). In general, students were expected to understand a range of concepts and laws organized around such domains as force and motion, electricity and magnetism, waves and optics, etc. Typically, understanding was demonstrated by the solution of problem types using quantitative methods that minimally demanded algebra. However, over the past several decades, research on student understanding has called into question whether the goals of instruction were being achieved.

The Route: Progression of Understanding

Historically, the implicit assumption in physics instruction has been that novice students could come to understand physics by receiving classroom presentations of what physics experts know. Research, however, has uncovered problems with that assumption. Students' naïve ideas and conceptions about the physical world are not easily changed and in fact often remain substantially unaffected by typical classroom instruction. In the 1970s and 1980s, research conducted by John Clement (1982), Andrea diSessa (1982), Lillian McDermott (1984), and others revealed that even those students who can recall physics laws and use them to solve textbook problems may not understand much about the implications of these ideas in the world around them. For example, in diSessa's research, college physics students performed no better than elementary students when asked to strike a moving object so that it will hit a target with minimum force at impact. Students relied on their untrained ideas in this task, ignoring the role of momentum, even when they could precisely reproduce the relevant laws of momentum on a test. Similarly, a study of student solutions to a problem with simple electrical circuits confirmed that students can reproduce scientific knowledge for a test, but revert to everyday ways of thinking when that knowledge is tapped outside the classroom (see Box 4.1).

Additional work over many years has led to the conclusion that students bring to physics a substantial set of persistent conceptions that are significantly different from those needed to understand aspects of the physics curriculum. By far, the largest amount of work on student conceptions has been in the area of Newtonian mechanics (McDermott and Redish, 1999). Both before and after passing high school and even college physics courses, students often behave as if their conceptual understandings of force and motion are more in line with pre-Newtonian, even Aristotelian conceptions. Although these ideas are misconceptions from a scientific perspective, they are sensible interpretations people construct from their everyday experience with objects and events in the world. Because these intuitions serve quite well to explain and predict many features of everyday phenomena, they are often unexamined and may be difficult to change. Yet instruction often proceeds as if students have no ideas at all—as if the job of teaching is simply to provide the ideas that are scientifically correct. Students' well-learned and practically useful mental representations for making sense of the world around them need to be actively engaged and built on if instruction is to be successful.

Thinking like a scientist is partly a matter of understanding how scientific principles are embodied in familiar events around us. This is a challenge for students. Like novices in any area, they are often too reliant on surface features when they attempt to interpret and solve physics problems. Students tend to look for clues in the objects featured in the problem. For example, Larkin (1983) found that novices often rely on the objects mentioned in the problem statement, like blocks, inclined planes, and pulleys, to try to construct a basic representation that consists of relations among these explicit objects. From this basic representation, they then seek directly to identify a set of equations that they can use to plug in the values mentioned in the problem.

In contrast, experts first construct an intermediate interpretation that represents the elements of the problem in constructs of the discipline, such as forces, acceleration, mass, momenta, etc. Chi and Bassok (1989) refer to this level of interpretation as a physics representation. They point out that the entities in a physics representation are not directly described but must be inferred. Perhaps because students tend not to identify problems as being members of categories defined by common scientific principles, they often fail to transfer what they "know" in the context of one problem to novel or even analogous problems.

Researchers have focused considerable effort on mapping out what students do understand about a variety of physical phenomena and how that understanding progresses as a consequence of instruction. This work has probed the conceptual understandings of learners from preschool onward. Minstrell's

BOX 4.1 Understanding Electrical Circuits

Eric Mazur gave two questions very similar to the following two problems to students in his Harvard algebra-based physics class.

Problem 1. Find the current through the 2 ohm resistor and the potential difference between points A and B.

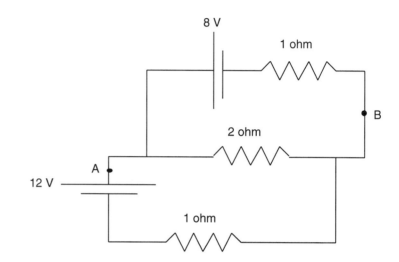

FIGURE 4.1a Diagram of a resistor.
SOURCE: Mazur, 1997.

The average score on the first (quantitative) question was 75 percent, while the average score on the second (qualitative question) was 40 percent.

Physicists consider the second question to be much simpler; in fact, they would consider parts of it to be so trivial and easy that many would not bother to give such a question on an examination. The correct reasoning is that closing the switch causes a short circuit across the third light bulb, reducing the total resistance in the circuit. With no potential difference across the third bulb, it has no current and goes out. With the same voltage applied by the battery and less total resistance, the voltage drop across each of bulbs (1) and (2) increases, the current through the battery increases, the brightness of bulbs (1) and (2) increases, and the power dissipated in the circuit increases.

work with high school students (e.g., Minstrell, 1989; Minstrell and Simpson, 1996), studies pursued by the Physics Education Group at the University of Washington with college students, and the research of many other investigators provide a wealth of information about how students typically think about a range of physical situations and concepts.

The question in Box 4.2 regarding the relative weight of an

Problem 2. In the circuit shown below, explain what will happen to the following variables when the switch is closed:

(a) the current through the battery
(b) the brightness of each of the bulbs
(c) the voltage drop across each of the bulbs
(d) the total power dissipated

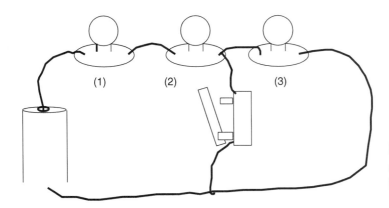

FIGURE 4.1b Diagram of a circuit.
SOURCE: Mazur, 1997.

The techniques that students applied to quantitatively solve the first problem could be easily applied to the second problem by assuming values for the various quantities and solving the problem quantitatively by comparing calculations for the circuit with the switch open with those for the circuit with the switch closed. Instead, many students give answers that are consistent with naive conceptions of how electric circuits work. Similar results have been found with thousands of students after traditional instruction in research done by the Physics Education Group at the University of Washington.

object when it is surrounded by air and submerged at two different depths of water produces a predictable range of responses from students when asked before, during, and even after instruction. Those responses can be evaluated for consistency with the various forms of student understanding shown in the box. The different answers, reflections of what Minstrell (1992) refers to as "facets" of thought, can be sequenced with

BOX 4.2 Understanding Fluid/Medium Effects and Gravitational Effects

A solid cylinder is hung by a long string from a spring scale. The reading on the scale shows that the cylinder weighs 1.0 lb.

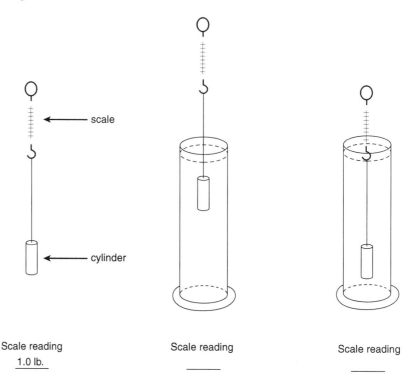

About how much will the scale read if the cylinder which weighs 1.0 lbs. is submerged just below the surface of the water? What will it read when the cylinder is much deeper in the water?

Briefly explain how you decided.

FIGURE 4.2 Sample constructed-response item: separating fluid/medium effects from gravitational effects.

• •

respect to scientific sophistication. They range from acceptable understandings in introductory physics (310) to those representing partial understanding (e.g., 315), to those representing more serious misunderstandings (e.g., 319).

Minstrell (1992) has argued that partially correct understandings frequently arise from formal instruction and may represent over- or undergeneralizations or misapplications of a student's knowledge. These can result if the set of examples

Forms of Student Understanding

310—pushes from above and below by a surrounding fluid medium lend a slight support (pretest—3 percent).

311—a mathematical formulaic approach (e.g., rho $\times$ g $\times$ h1 rho $\times$ g $\times$ h2 = net buoyant pressure).

314—surrounding fluids don't exert any forces or pushes on objects.

315—surrounding fluids exert equal pushes all around an object (pretest—35 percent).

316—whichever surface has greater amount of fluid above or below the object has the greater push by the fluid on the surface.

317—fluid mediums exert an upward push only (pretest—13 percent).

318—surrounding fluid mediums exert a net downward push (pretest—29 percent).

319—weight of an object is directly proportional to medium pressure on it (pretest—20 percent).

presented to students is too limited or if an appropriate set of contrasting cases is not included to help clarify the conditions under which a concept applies. The task for a student is to come to recognize similar situations and problems as members of a category. Part of the challenge, then, is to understand the range of conditions under which concepts apply. It is important for both the instructor and the student to become aware of the form of the student's conceptual understanding when instruction begins and to monitor changes in the student's knowledge as instruction progresses. The goal is to build on what students do know and to help them understand the conditions under which it applies, rather than ignoring students' current concepts and trying to replace them immediately with scientific reasoning.

The Vehicle: Pedagogy and Curriculum

Along with the research on student understanding have come new approaches to teaching physics that clearly demonstrate the accessibility of the subject for all students, if it is taught in ways that acknowledge what is known about student understanding.

First, instruction needs to be based on the acknowledgment that students are being asked to reformulate category systems that have served them quite well in the past. This entails coming to recognize apparently familiar objects and events as members of novel classes, an accomplishment that develops slowly and only if students receive multiple well-chosen opportunities to experience the relevant range of situations in which a concept applies or does not apply. Second, teachers need to be aware of the range of ways in which students interpret situations and problems and to develop a repertoire of proven strategies for helping them question their assumptions, tune their partly correct conceptions, and understand the boundary conditions for important principles. The goal is to help students understand, which requires knowing how to capitalize on the forms of sense-making that they have available to work with. Third, effective physics instruction is designed to make more transparent what the practice of physics is all about. As Hestenes (1987) has cogently argued, physics is a "modeling game," but this is far from apparent to students. The emphasis in much physics instruction is on using the products of physics—laws and principles—with little attention to why or how physicists generate and work with these concepts. Students rarely are taught physics as the enterprise of constructing, testing, and revising models of the world, and therefore its primary goals and epistemology are typically invisible. As we explain below, however, new approaches to instruction are emphasizing this aspect, inviting students into the modeling game and making evident the goal of what otherwise may seem a rather mysterious enterprise.

To summarize, we emphasize three characteristics of new

approaches to physics instruction that follow from contemporary learning research:

1. helping students develop new schemas for recognizing objects and events in the world as members of a category system more like the one used by scientists;
2. continually monitoring changes in students' evolving knowledge to inform choices of just the right experiences, countercases, and challenges to support their next step in knowledge development; and
3. introducing students to physics as a modeling game, so that they grasp its epistemology and central goals.

Most of the more fully developed curricula inspired by this research have been targeted at the college level. McDermott and Redish (1999) have identified nine such curricula and more are under development, including a research-based version of the widely used college text by Halliday and Resnick (Cummings et al., 2001). However, the substantial overlap of introductory college and high school physics courses suggests that much in these curricula may also be appropriate for high school use.

Two programs designed specifically for students in middle school and high school have demonstrated improvements in student achievement, particularly with respect to conceptual understanding. In the first of these, the "modeling method" of instruction, students work to develop, evaluate, and apply their own models of the physical behavior of objects (see Box 4.3). The key to this instructional intervention is a series of professional development workshops with teachers, who are supported in effecting a radical shift of their pedagogy. Teachers are encouraged to become modeling coaches, helping students to observe, model, and explain interesting and puzzling phenomena. A 6-year project that provided extensive training and support for 200 teachers in this instructional approach resulted in nearly all of them demonstrating significant improvements in their own understanding, in their teaching, and in their students' achievement (Hestenes, 2000; for more information, see http://modeling.la.asu.edu/modeling.html) on a highly regarded measure of physics conceptual understanding, the Force Concept Inventory (see below). These studies, conducted with large numbers of students in matched comparison groups, were carried out in multiple sites across several years. The research included students in regular and introductory physics classes, honors-level physics, and advanced placement physics. Results repeatedly showed greater pretest to posttest gains in physics content knowledge when students taught by the modeling method were compared with (a) physics students of the same teachers in the year before the teachers implemented the program and (b) students in traditional physics classes and alternative reform programs. Students taught with the modeling method exceeded the performance of comparison groups by

- -

BOX 4.3 Modeling Instruction in High School Physics

The modeling method has been developed to address problems with the fragmentation of knowledge in traditional physics instruction and the persistence of naive beliefs about the physical world. It is an approach to high school physics instruction that organizes course content around a small number of basic scientific models as units of coherently structured knowledge. David Hestenes and colleages at the University of Arizona have developed the approach to both instruction and teacher preparation over the past two decades. The program is grounded on the thesis that scientific activity is centered on modeling: the construction, validation, and application of conceptual models to understand and organize the physical world.

Instructional activities give students experience in constructing and using models to make sense of a variety of physical problems. A critical feature of the program is the role played by the teacher: "The teacher cultivates student understanding of models and modeling in science by engaging students continually in 'model-centered discourse' and presentations." The program developers argue that "the most important factor in student learning by the modeling method (partly measured by Force Concept Inventory scores) is the teacher's skill in managing classroom discourse" (Hestenes, 2000, p. 2). The teacher is prepared with a definite agenda for student progress and guides student inquiry and discussion in that direction with Socratic questioning and remarks.

The program uses computer models and modeling to develop the content and pedagogical knowledge of high school physics teachers, and relying heavily on professional development workshops to equip teachers with a teaching methodology. Teachers are trained to develop student abilities to make sense of physical experience, understand scientific claims, articulate coherent opinions of their own, and evaluate evidence in support of justified belief. Teachers are also equipped with a taxonomy of typical student misconceptions in order to prepare them to identify and work with them as they surface.

In a sample of 20,000 high school students, gains on the Force Concept Inventory under modeling instruction are reported to be on average double those under traditional instruction, with teachers who implement the program more fully showing higher gains in their students scores. All students gained significantly from modeling instruction, but students with the lowest scores before instruction gained most.

More information on the modeling method is available on the projects web site: http://modeling.la.asu.edu/modeling.html.

margins that in some cases were larger than two standard deviations.

The second program, ThinkerTools, also emphasizes physics as modeling and features both computer simulations and inquiry with physical materials. The curriculum (White, 1993; White and Frederiksen, 1998) teaches the physics of force and motion and is designed to be successful with students in middle school and in urban environments as well as in high school and in suburban environments.

The inquiry-based curriculum engages students' preconceptions by asking them what they think will happen when certain forces are applied to objects. Students test their ideas in a computer-simulated world and learn when their ideas hold true and when not. The observations allow students to directly challenge their ideas and to engage in a search for a theory that can adequately explain what they observe. The class functions as a research community, and students propose competing theories. They test their theories by working in groups to design and carry out experiments using both computer models and real-world materials. Finally, students come together to compare their findings and to try to reach consensus about the laws and causal models that best account for their observations.

Students systematically build conceptual understanding by encountering problems that increase in complexity and difficulty. The problems are based on knowledge about typical forms of student thinking and its progression. Experiences that students encounter support the "conditionalized" kind of knowledge that experts hold, allowing students to detect patterns that novices do not see. ThinkerTools provides multiple experiences with problem solving, but the carefully controlled difficulty of problems is designed to build pattern recognition efficiently.

Like the modeling method, the emphasis is on constructing and revising models and explanations, and modeling ability is acquired in the service of building a conceptual understanding of motion, gravity, friction, and the like. What is distinctive in the ThinkerTools curriculum is the addition of a "reflective assessment" component. In addition to engaging in inquiry learning, students learn to evaluate the quality of their own and others' inquiry investigations using standards that reflect the culture and the goals of the scientific community.

As with the modeling method, student achievement gains with the ThinkerTools curriculum are impressive. Students construct a deeper understanding and are better able to transfer their knowledge to novel problems than students taught with a traditional curriculum. These advantages held even when the ThinkerTools group consisted of urban students who were compared with suburban counterparts, as well as middle school students compared with high school students. Of particular importance is the finding that teaching students to engage in reflective assessment—to judge how well they and their colleagues carried out an inquiry investigation—substantially improved learning gains. Not only did students come away with a deeper understanding of the inquiry process, but they also improved their content knowledge of physics. The gains were particularly striking for students who began instruction as low achievers (see Box 4.4).

Checkpoints: Assessment

New forms of instruction pose two fundamental challenges for assessment. First, since the goal of instruction emphasizes conceptual understanding and the ability to transfer knowledge to new situations, assessments are needed that capture these competences. Second, since the instructional goal is to support conceptual change, instructors need assessment tools to continually diagnose student thinking so that instruction can address students' evolving conceptions.

With respect to the first challenge, physics teachers and researchers alike are increasingly adopting the Force Concept Inventory (see Box 4.5) developed initially by Halloun and Hestenes (1985) and later modified and published with comparison data (Hestenes et al., 1992). This instrument probes beyond the usual focus on students' capability to solve traditional physics problems, emphasizing instead their conceptual analysis of physical situations. Rather than right or wrong answers, the inventory diagnoses student conceptions; the items and choices in the instrument are based on research about the range of thinking that students typically bring to situations featured in the test. Because it provides feedback about students' conceptual development, it has persuaded some instructors of the need to make significant changes in their teaching (see, e.g., Mazur, 1997). Several other tests modeled after the Force Concept Inventory are currently available or under development in areas of physics beyond force and motion. One example is the conceptual survey of electricity and magnetism described by Maloney et al. (2001).

Although it is certainly useful to know how students think about physical situations and concepts at the start and end of instruction, it is also important to monitor changes in these ideas during the course of instruction to address specific student needs and modify instruction accordingly. Incorporating formative assessment practices into ongoing instruction requires quality assessment materials that are closely connected to conceptual models of student understanding, together with effective ways of presenting, scoring, and interpreting the assessment results. Time and efficiency are obviously of central importance: if the feedback is not available when the next instructional decisions need to be made, then important opportunities will be lost. An excellent example of an effort to integrate assessment and instruction is the facets-based instruction and assessment work performed by Minstrell and his colleagues described above (Minstrell, 1992; Hunt and Minstrell, 1994; Levidow et al., 1991). The focus of the research effort has been on identifying facets (mental representations for interpreting physical situations) of student knowledge and understanding for various topics in physics. These facets are incorporated into Diagnoser, a relatively simple-to-use computer program designed to help teachers evaluate the quality and consistency of student reasoning in physical situations.

The Diagnoser program presents sets of carefully designed problems and records student responses and justifications as a means of identifying their understanding. When needed, the program provides instructional prescriptions that are designed to challenge the student's thinking and address a possible conceptual misunderstanding. The course instructor is provided with information about the range of student understanding in the class and can then adjust lessons accordingly. Minstrell and Hunt (1990) have demonstrated that the facets approach can be successfully adopted by teachers and that it produces better outcomes than instruction that lacks integrated diagnostic assessment.

TEACHER KNOWLEDGE

Teacher knowledge of physics is typically not a serious concern, since most physics teachers have undergraduate or ad-

BOX 4.4 Reflective Assessment in ThinkerTools

ThinkerTools is an inquiry-based curriculum that allows students to explore the physics of motion. The curriculum is designed to engage students' conceptions, to provide a carefully structured and highly supported computer environment for testing those conceptions, and to steep students in the processes of scientific inquiry. The curriculum has demonstrated impressive gains in students' conceptual understanding and ability to transfer knowledge to novel problems.

White and Frederiksen (1998) designed and tested a reflective assessment component that provides students with a framework for evaluating the quality of an inquiry investigation—their own and others. The assessment categories included understanding the main ideas, understanding the inquiry process, being inventive, being systematic, reasoning carefully, using the tools of research, teamwork, and communicating well. The performance of students who were engaged in reflective assessment was compared with that of matched control students who were taught with ThinkerTools but were asked to comment on what they did and did not like about the curriculum without a guiding framework. Each teacher's classes were evenly divided between the two treatments. There were no significant differences in students' initial average standardized test scores (the Comprehensive Test of Basic Skills was used as a measure of prior achievement) between the classes assigned (randomly) to the different treatments.

Students in the reflective assessment classes showed higher gains both in understanding the process of scientific inquiry and in understanding the physics content. For example, one of the outcome measures was a written inquiry assessment that was given both before and after the ThinkerTools inquiry curriculum was administered. It was a written test in which students were asked to explain how they would investigate a specific research question: "What is the relationship between the weight of an object and the effect that sliding friction has on its motion?" (White and Frederiksen, 2000:22). Students were instructed to propose competing hypotheses, design an experiment (on paper) to test the hypotheses, and pretend to carry out the experiment, making up data. They were then asked to use the data they generated to reason and draw conclusions about their initial hypotheses.

Presented below are the gain scores on this challenging assessment for both low- and high-achieving students and for students in the reflective assessment and control classes. Note first that students in the reflective assessment classes gained more on this inquiry assessment. That this was particularly true for the low-achieving students. This is evidence that the metacognitive reflective assessment process is beneficial, particularly for academically disadvantaged students.

This finding was further explored by examining the gain scores for each component of the inquiry test. As shown in the figure below, one can see that the effect of reflective assessment is greatest for the more difficult aspects of the test: making up results, analyzing those results, and relating them back to the original hypotheses. In fact, the largest difference in the gain scores is that for a measure termed "coherence," which reflects the extent to which the experiments the students designed addressed their hypotheses, their made-up results related to their experiments, their conclusions followed from their results, and their conclusions were related back to their original hypotheses. The researchers note that this kind of overall coherence is a particularly important indication of sophistication in inquiry.

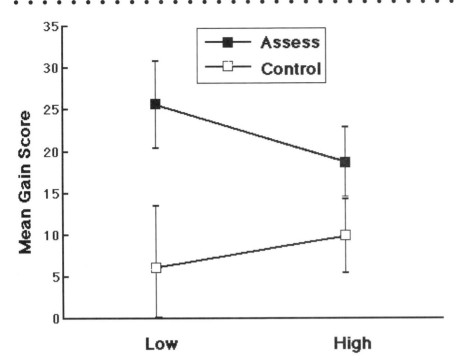

FIGURE 4.4a CTBS achievement levels.
SOURCE: White and Frederickson, 2000.

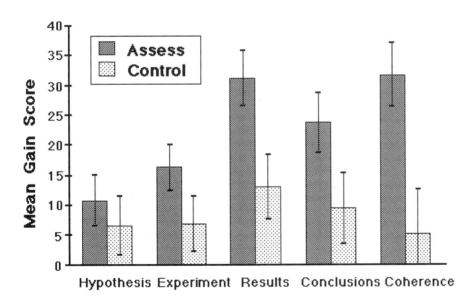

FIGURE 4.4b Average gains on the inquiry test subscores for students in the reflective-assessment and control classes.
SOURCE: White and Frederickson, 2000

BOX 4.5 The Force Concept Inventory

The Force Concept Inventory (Hestenes et al., 1992) has been widely used to compare student mastery of basic concepts of Newtonian mechanics. The test examines core conceptual understanding of Newtonian mechanics.

Sample Question

Imagine a head-on collision between a large truck and a small compact car. During the collision:

(A) The truck exerts a greater amount of force on the car than the car exerts on the truck.
(B) The car exerts a greater amount of force on the truck than the truck exerts on the car.
(C) Neither exerts a force on the other; the car gets smashed simply because it gets in the way of the truck.
(D) The truck exerts a force on the car but the car doesn't exert a force on the truck.
(E) The truck exerts the same amount of force on the car as the car exerts on the truck.

Correct answer is (E).

This question assesses student understanding of Newton's third law. The distractors (incorrect answers) are adapted from student responses in interviews and open-form questions, revealing various naive conceptions of force as associated with size or effect. Newtonian principles demonstrate that force is an interaction, so the forces are the same, but the effects of the forces (acceleration, damage) differ according to the mass and structure of the object.

The inventory is commonly given in a pretest-posttest mode. It is inconceivable to most teachers that a student well trained in mechanics could do poorly on these core concepts on the posttest. Most physics teachers agree that a student with a reasonable understanding of Newtonian mechanics should be able to correctly answer the 30 simple questions on the test, such as the one illustrated above. Indeed, the test seemed so simple that many instructors initially did not think it was worth administering as either a pretest or a posttest. Yet students do poorly on the inventory as a pretest, and a full semester of careful traditional instruction produces little change in student performance: this result has been a major wakeup call to many physics teachers. Such results, which teachers can often replicate with their own classes, have significantly increased the audience for the results of physics education research.

vanced physics training. Of far greater concern is teachers' pedagogical content knowledge. As the above discussion suggests, the knowledge and tools are now available to support the latter, including the research base concerning students' conceptual understanding, assessment tools such as the Force Concept Inventory, and alternative forms of instruction, such as the emphasis on modeling described earlier. It is uncertain what proportion of the 19,000 physics teachers in the United States engage in instructional practices that are aligned with what is known about learning and instruction in physics, or how many make use of research-based curricular materials, assessments, and approaches. Equally uncertain is the source of their knowledge, that is, whether generalized from their own experiences as physics students, acquired during pre-service teacher education, or developed as a result of professional development programs.

In any given year, the number of physics majors pursuing a secondary education teaching credential is relatively small, and in most institutions of higher education only a few of these students may be simultaneously pursuing a certification program. Variation in program content, student learning experiences, and supervision can be substantial. This is especially problematic with regard to the specifics of how prospective physics teachers acquire knowledge about important characteristics of student learning and the teaching of physics.

In contrast to pre-service teacher education, the professional development of in-service physics teachers is often better organized, especially with regard to regional, state, and national workshops. Many of these opportunities have been supported by federal funding such as the Eisenhower math-science programs, the National Science Foundation's teacher enhancement projects. Professional societies have played a role as well. The physics teaching resource agent (PTRA) program, run by the American Association of Physics Teachers, is one model of a sustained teacher enhancement project. First funded in 1985, the PTRA program develops workshop materials, prepares exemplary high school teachers to serve as resource agents, and provides support to those agents to offer workshops in their own regions (Nelson and Bader, 2001). Agents continue to receive education over successive summers to expand their repertoire of workshops.

Approximately 500 outreach teachers have been educated, and more than 300 remain active. From 1985 to 1995, the outreach program offered workshops to about 60,000 teachers, including physics, physical science, middle school, and elementary school science teachers. Since 1995, the project has continued with the urban PTRA project and the rural PTRA project. In these continuations, there is an emphasis on building a continuing relationship with a group of teachers for sustained professional development. The PTRA now serves as a model for the development of outreach programs in other science areas.

The opportunities for sustained teacher learning are better developed in physics than in most fields. Yet even here, relatively little is known about the processes of effective teacher learning. In physics as elsewhere, little is understood about how knowledge of student thinking is bound to practice—that is, how it is used by teachers to deploy specific instructional moves. Little is known about the conditions that optimize and impede the development of the knowledge base of prospective and new teachers or the conditions that assist experienced teachers in understanding and adopting more effective instructional practices. Finally, little is known about the range of teacher characteristics and organizational circumstances that are conducive to adopting the instructional approaches derived from physics education research (like the modeling method or facets-based instruction). Because the knowledge base and tools for teacher learning are better developed in physics than in other areas, it provides fertile ground for investigating these broadly applicable questions.

RESEARCH AGENDA

Three closely related initiatives for research and development are proposed that would build on the existing, high-quality research and development in physics and support the usefulness of existing knowledge and tools for school decision makers and classroom teachers:

1. undertake an effort to identify and differentiate existing research-based physics instruction programs on dimensions of learning outcomes and characteristics of students, teachers, and schools in which the program has been effective;
2. study the scalability of the existing programs and develop supports for taking programs to scale;

3. study teacher knowledge requirements for effective use, and how that knowledge builds with teacher learning opportunities and experience.

Initiative 1: Differentiating Instructional Programs and Outcomes

The initiative should begin by developing a better characterization of existing programs, as well as the range and scope of their use, for purposes of informing education decision makers. One set of questions concerns what conditions usually accompany success: participation from university or other research partners; electronic or physical proximity to a network of more expert reform teachers; administrative support and resources; aligned policies about assessment, grading, and student promotion.

The analysis should also identify the dimensions on which these research-based programs differ from each other and how the differences affect outcomes. Some of the programs rely heavily on computer simulations and data-gathering tools, and others do not. Some emphasize student reflection more than others. Some are targeted primarily toward students, whereas others are targeted primarily toward changing the practices of teachers. A useful first step, then, would be to provide a systematic catalogue of the current state and scope of available programs, evidence about their outcomes, and analysis of the features that may account for variability in outcomes.

These kinds of comparisons are often difficult to make because the research designs, measures of success, and data collected differs from one program evaluation to the next. A likely required step in this effort would be the design of research to allow the questions regarding relative effectiveness under varying conditions to be answered more robustly. The investigations would be strengthened considerably if researchers work in schools and conduct cycles of repetition-with-variation of instruction so that critical variations in program features can be explored.

Initiative 2: Scalabilty

In spite of the promising nature of the findings to date, only a small fraction of physics classes currently use research-based programs. We therefore propose exploring the extent to which promising programs can be taken to scale. This work could be undertaken in a set of SERP field sites that exhibit a range of student, teacher, and school characteristics. Careful study of variation in the success of the program, as well as the characteristics and conditions that can explain that variation, would be the primary research task.

As instructional programs move farther away from the sites and the individuals who generated them, they typically undergo change. As Spillane (2000) and others have shown, reform curricula may look far different in practice than what was intended in the reform design. Very often, policy makers and practitioners generate piecemeal interpretations of new approaches to instruction, preserving some of its surface features while missing its underlying goals—rather like the physics novices described earlier. So, for example, a teacher initially encountering an unfamiliar way to teach physics might incorrectly conclude that the most important feature of a research-based curriculum is its hands-on approach. Of course, students can have their hands on many things, often in ways that fail to promote any meaningful conceptual development. The point is that the more new programs and curricula deviate from currently familiar practices, the more likely they are to be distorted or misunderstood. These misinterpretations, when combined with the need to adapt to local circumstances, can lead to wide variability in the program in practice.

Consequently, instructional programs that successfully build on knowledge from research and that demonstrate effectiveness in experimental studies can show insignificant, or even negative, results when implemented more broadly. The very notion that research can improve practice is undermined by the outcome. If research is to have a positive, widespread impact on student learning, following curriculum use as it spreads into school districts and is adapted by teachers will be critical.

The envisioned work would seek to identify the variability of implementation in the new programs. Are there programs in which fidelity is generally preserved and others that are more frequently distorted? If so, which programmatic features are responsible? Are there adaptations that improve program performance?

These investigations require the kind of longer term study that SERP can support. Moreover, it will be important to learn about the forms and amounts of variability in implementation that programs can sustain (that is, program robustness), and the resulting outcomes that can be expected from typical variations in enactment.

The research effort should be accompanied by an effort to develop supports for effective implementation. If, for example, a specific feature of a program is easily misunderstood, the program itself may need to be revised, or supplemental opportunities for understanding may need to be developed for users. Multiple iterations of research, design, and testing will be required to develop supports that are effective.

Initiative 3: Teacher Learning

Much more research has been devoted to understanding student learning of physics and how to improve it than has been spent on corresponding studies of teacher learning and how to improve it. As yet, we know very little about how effective physics teachers use their knowledge of student thinking to make decisions about which move in their instructional repertoire to deploy in a situation. Some clues do exist. For example, Minstrell has formalized an expert teacher's diagnostic knowledge into his assessment program, Diagnoser (discussed above). But the instructional move an expert might make to respond to a given facet identified by Diagnoser is still poorly understood. Nor do we have a good sense of how this kind of knowledge develops—that is, at what characteristic rates and under what conditions.

Physics is an excellent topic for supporting a SERP study of the development of teacher knowledge, because so much of the student diagnostic work has already been accomplished. And successful efforts to educate teachers, like that employed by the modeling instruction program described above, provide fertile ground for research. SERP could support longitudinal studies to generate a richer sense of the typical development of teacher knowledge, from pre-service student, through novice, to journeyman teacher. These studies might also help us understand what hampers the continuing development of this form of knowledge and its connection to instructional decision making. On one hand, because student conceptions in physics are so robust, it seems unlikely that novice teachers are entirely unaware that their students hold them. On the other hand, neither teachers' education nor the forms of teaching they adopt may provide the kinds of feedback that help them develop a systematic catalogue of those conceptions and a repertoire of productive approaches to addressing them. Experiences that provide this kind of information at the appropriate level of detail to guide instruction are particularly important to understand. This research would clearly support efforts to understand teacher knowledge requirements in other subject matter, with very close parallels, for example, to the teacher knowledge agenda in reading comprehension.

SCIENCE EDUCATION ACROSS THE SCHOOL YEARS

It would surely be disturbing if the mathematics instruction in schools followed no plan for increasing students' knowledge cumulatively across grades of study but instead meandered from topic to topic in an unprincipled way. Yet this is an accurate description of science instruction in elementary schools and in many middle schools. High schools have a more predictable sequence of science subjects rooted in tradition, but the subjects are generally treated separately. Even in high school, there is little effort devoted to drawing connections in the content across subjects or to systematically building an understanding of the discipline.

STUDENT KNOWLEDGE

The Destination: What Should Students Know and Be Able to Do?

Achieving consensus on the content of science education across the K-12 school years has been stymied by a long history of debates and subsequent confusion about the appropriate organizing principles for science education. The debates often swell around the process-content divide. Some have argued that the most important thing for students to learn is the process of scientific reasoning, including the logic of controlling extraneous variables in scientific experimentation, the coordination of theory and evidence, and standards for evaluating evidence. However, these attempts often founder on superficial and fragmented treatments of science content. Too exclusive an emphasis on scientific processes can result in instruction about a bundle of topics that are loosely, if at all, related to each other because the development of scientific reasoning does not depend on the treatment of particular topics. Units on weather, electricity and magnetism, and the rain forest follow each other, in an organization most charitably described as modular. Knowledge accumulated in early grades does not build smoothly toward the scientific ideas that will be encountered in high school study and beyond.

In contrast, those who endorse a content approach seek coherence by emphasizing the integrated development of knowledge within scientific disciplines, like biology or chemistry. In practice this emphasis often leads to a focus on concepts and facts—the products of science—with little attention to how that knowledge was generated. In earlier grades, students receive instruction that jumps from earth sciences to physical sciences to biological sciences. The usual result is superficial or fragmentary understanding (Vosniadou and Brewer, 1989; Pfundt and Duit, 1991).

Neither of these views is well aligned with the vision sketched in national science standards (e.g., those from the American Association for the Advancement of Science [AAAS] and the National Research Council). The standards point toward the big ideas and themes that ought to be the goals of science education. For example, AAAS's report (1991) *Science for All Americans* proposes that by the time they graduate, students should understand important scientific themes like systems, models, constancy and change, and scale. AAAS points out that these themes "transcend disciplinary boundaries and prove fruitful in explanation, in theory, in observation, and in design" (p. 165). However, there are few illustrations in practice of what it means to understand these ideas deeply, and few guideposts to help teachers navigate the very extensive list of topics that the standards include so that students will arrive at deep understanding of these themes or organizing big ideas. Overall, very little is known about how the material outlined in the standards is actually attainable over the time course of schooling.

The Route: Progression of Understanding

From a very young age, children begin to impose order on the world they observe, generating ideas about why objects float or sink, about what it means to be alive, about why plants grow, and about temperature and weather. Like the high school physics students described in the preceding section, children are sense-makers, and their theories and ideas are tools that should be encouraged and stretched, not ignored. As children accumulate greater experience with the world, they begin to master the kinds of distinctions that adults and scientists make: dogs are indeed alive and rocks are not, but plants are also alive, even though they do not move from place to place of their own volition.

Early conceptions may evolve somewhat with increased experience. But *scientific* conceptions generally do not develop without explicit instruction, partly because the epistemological assumptions that underlie them are complex and often invisible. As discussed with regard to physics, many important scientific processes, principles, and laws are at odds with everyday understandings and experiences. Much scientific work requires methods and measurements that are not characteristic of everyday activity, and that rely on instruments that allow the scientist to explore what is not otherwise accessible.

Even more fundamentally, it is by no means self-evident to children what kind of enterprise science is. Experimentation, for example, requires arranging aspects of the world to generate a model of the phenomenon that is of interest. The model, which is taken to stand for the more general class of events, is then systematically perturbed as a way to seek deeper understanding. The history of science suggests that this way of constructing knowledge evolved gradually, as practicing scientists increasingly came to regard their work as the creation of a form of argument, rather than as the unproblematic observation of transparent events in the world (e.g., Bazerman, 1988).

While a scientific approach is not likely to develop in children naturally, this form of thinking can be developed gradually when students have sustained opportunities (as in Sister Hennessey's classroom, described below) to learn the discipline's content knowledge (what we know), theory (what we make of what we know and don't know), and knowledge of the epistemologies of science (how we know). But little is known about the routes or the progression of understanding that characterize effective mastery of science content and reasoning.

What science are children capable of learning at different grade levels? Elementary school children studying marine mammals may be quite capable of understanding that the ancestors of whales lived on land, and they may also be engaged by the story of how that came to be known by scientists. But less is known about their readiness to understand the concepts of distribution and variation that underlie such evolutionary tales. Some evidence suggests that even elementary and middle school students can begin to develop an understanding of these ideas (see, for example, Cobb et al., 2003; Lehrer and Schauble, 2001, 2002; Petrosino et al., 2003). But little systematic research has been done to discern what the majority of children are able to grasp with reasonable instructional effort at different grade levels. Furthermore, we know little about what instruction is required at one level to prepare students for instruction at the next. Most instructional research is conducted over brief periods of time and so does not provide information about the potential for long-term development of knowledge and reasoning.

AAAS Project 2061 published a set of science literacy maps that lay out a progression over grades in the components of knowledge that students should develop for each of the AAAS benchmarks (American Association for the Advancement of Science, 2001). At present, this atlas represents the only comprehensive attempt to establish a developmental course of learning and instruction for students in grades K-12. But it, like the National Research Council science standards and the AAAS benchmarks, lacks a research foundation to support the assumptions about learning and the progression of understanding. They are conjectures that, while reasonable, lack empirical confirmation. Moreover, since they are based on a sense of the way that "typical" children think (that is, under no particular conditions of instruction), they are very likely to be underestimates of children's capabilities. Also missing is knowledge about how to provide appropriate sequences of instruction, as well as a clear sense of the ways in which the standards and benchmarks map against assessments of students' knowledge representations and cognitive skills.

The Vehicle: Curriculum and Pedagogy

The knowledge base to support the development of curriculum and pedagogy, we have argued, is characterized by little detail on the instructional implications of teaching the big ideas and little understanding of the progression of student thinking that is possible with instruction both in science content and process. Those weaknesses in the knowledge base are reflected in the K-12 science curriculum.

Analyses of TIMSS science achievement results (Schmidt, 2001; Valverde and Schmidt, 1997) as well as research conducted by other investigators show that in contrast to other countries, elementary and middle school science in the United States emphasizes broad coverage of diverse topics over conceptual development and depth of understanding. For example, eighth grade textbooks in the United States cover an average of more than 65 science topics, in stark contrast to the 25 topics typical of other TIMSS countries. "U.S. eighth-grade science textbooks were 700 or more pages long, hardbound, and resembled encyclopedia volumes. By contrast, many other countries' textbooks were paperbacks with less than 200 pages" (Valverde and Schmidt, 1997:3). The more recent TIMSS-R follow-up study concluded that the comparatively poor performance of U.S. eighth graders is related to a middle school curriculum that is not coherent and is not as demanding as that found in other countries studied. "We have learned from TIMSS that what is in the curriculum is what children learn" (Schmidt, 2001:1).

Commercially published textbooks are the predominant instructional materials used in science (Weiss et al., 2002). In grades K-4, textbooks are used 65 percent of the time; this increases to 85 percent of the time in grades 5-8, and 96 percent of the time in grades 9-12. Most of these textbooks are seriously flawed. A team at the AAAS reviewed widely used textbooks in middle and high school science and ranked them on a number of criteria, among them the extent to which the major concepts were communicated clearly and students' preconceptions were addressed. All of the middle school textbooks and most of the high school textbooks were rated poor (Roseman et al., 1999). On the critical dimension of supporting conceptual change, widely used science textbooks at both the middle school and high school level have been judged poor by the AAAS team (Roseman et al., 1999).

In recent years, researcher-practitioner collaborations have begun to generate more systematic approaches to science education. These instructional efforts span several grades and are, in a sense, hypotheses about what children can learn and do at different grade levels. The commitments and design trade-offs vary somewhat from program to program, but they share the following features:

1. reformulation of the goals and purposes of school science;
2. including less content material at greater depth and an emphasis on understanding over coverage;
3. fostering and studying the long-term development of student knowledge under optimal conditions of teaching and learning;
4. sequencing curriculum on the basis of what is being learned about the development of student knowledge; and
5. conducting research on what it takes for these forms of teaching and learning to take hold and flourish in new settings.

These programs share the conviction that the central goal of science education should be to develop students' understanding and appreciation of the forms of knowledge-making that characterize scientific practice. At the same time, however, this goal is pursued in the course of serious and extended investigation of science content. While the program developers have been collecting data on learning outcomes that show promise, these programs have not undergone rigorous, independent testing. Their value to a SERP research agenda lies not in the definitive answers to instructional questions provided by the programs, but in the opportunity they provide to further develop and rigorously test hypotheses about alternative approaches to teaching science to young children. The programs can be differentiated as having one of four foci: methods of empirical inquiry and inference; theory building; modeling; or argumentation.

Methods of Empirical Inquiry Kathleen Metz (2000) at the University of California, Berkeley, has been working for several years with a cadre of elementary grade teachers to reorganize science teaching and learning around the enterprise of (and methodologies for) empirical inquiry (see Box 4.6). Students learn how to pose questions, to think carefully about how those questions could be answered empirically, and to master a repertoire of methods to conduct empirical investigations. Metz is generating longitudinal data, both about students' evolving content understanding (e.g., concepts about animal behavior, adaptation) and their capabilities to pose interesting questions and investigate them via studies of their own design, as well as to diagnose weaknesses in their own and other students' investigations.

• •

BOX 4.6 Science as Knowledge-Rich Goal-Focused Inquiry

Kathleen Metz has worked with teachers in grades 1 through 5 to engage children in authentic, goal-focused scientific inquiry. As students study a domain in greater depth, they are given increasing responsibility for the inquiry. The program grew out of Metz's concerns that existing curricula for elementary science assume developmental constraints on the students' ability to learn science at an early age that are not supported by research. As a result, the content of elementary science curricula is frequently impoverished, and the focus on discrete "science process skills" in many curricula is divorced from the robust context of inquiry that can enrich students learning opportunities and interest. Metz's instructional model uses a combination of empirical investigations, text, video, and case studies to develop content knowledge of the domain, and to introduce students to the big ideas of biology, process knowledge of tools and decision making involved in scientific inquiry, and science as a way of knowing. She has instantiated this instructional model in curriculum prototypes in animal behavior and botany.

For example, in the study of animal behavior, young children's observations of animals are used as a basis to discuss distinctions between observation and inference that children usually conflate into one undifferentiated category of "the way things are." Eventually, the children develop parallel distinctions between theory and evidence through case study, empirical investigations, and text analysis.

In one case, children take on the question animal behaviorist Roger Payne posed to himself of why grey whales migrate. They examine Payne's analysis of the theories he considered and his evidence for and against them. Analysis of text is also used to deepen their knowledge and support the development of individual interests and questions. The study introduces theory and evidence, in conjunction with the key idea of survival advantage. In another case, across a series of increasingly complex investigations of cricket behavior, the teacher supports the development of children's emergent repertoires of methods, ways to analyze data, and ways to represent data in the form of three "menus."

As students become more familiar with the inquiry process, they work in pairs to formulate the question they want to research, develop a plan to investigate the question, implement their study, and represent their work in the form of a research poster. In a research poster conference, the children analyze the surprises, the relation between their studies, and consider next steps in their research as a whole. Metz's bet is that scaffolding the content and process knowledge necessary for children to assume control of their *own* investigation has significant pay-offs from cognitive, epistemological, and motivational perspectives.

LEARNING AND INSTRUCTION

Theory Building An instructional approach designed by Sister Mary Gertrude Hennessey emphasizes theory building (see Box 4.7). Across content domains through the elementary years, students repeatedly consider the criteria by which scientific theories are formulated, used, tested, and revised. Sister Hennessey has generated longitudinal data concerning changes in her students' grasp and application of these criteria, and she has also tracked changes in students' propensities to reflect about their own thinking. Researchers external to the project have documented impressive performances by these students on standardized interviews concerning the nature of science.

Modeling A number of investigators are examining the potential of organizing science instruction around the practice of modeling. This kind of instruction emphasizes developing models of phenomena in the world, testing and revising models to bring them into better accord with observations and data, and, over time, developing a repertoire of powerful models that can

· ·

BOX 4.7 Science as Theory Building

Until recently, Sister Mary Gertrude Hennessey, who has Ph.D.s in both science and science education, served as the sole science teacher for students in Grades 1-6 at St. Ann School in Stoughton, Wisconsin (she is now serving as principal). "Hennessey's curricular approach stands out as an extensive and sustained attempt to teach elementary science from a coherent, constructivist perspective" (Smith et al., 2000:359). Hennessey's instruction emphasized theory-building, both as the process by which students build their own science understanding and as an object of explicit reflection. Across content domains, students repeatedly considered the criteria by which scientific theories are formulated, used, tested, and revised. This emphasis was consistently maintained across grades of study.

In early grades, the focus was on identifying and explicitly stating one's own beliefs and the alternative beliefs held by classmates. In later grades, Hennessey "raised the ante" by urging students to consider the advantages of adopting additional criteria, such as the intelligibility, plausibility, and extensibility of their beliefs and the beliefs of others. In every case, these issues were explored in the context of sustained investigations of phenomena. Students applied these criteria as they worked toward building deep explanations based on theoretical entities, investigating the implications of their own explanations and alternative explanations proposed by the class.

Sixth graders who spent six years under Hennessey's tutelage showed impressive epistemological development on the nature of science interview developed by Carey and colleagues (Carey et al., 1989; Smith et al., 2000).

be brought to bear on novel problems. Modeling approaches have the advantage of avoiding the content-process debates that have plagued science education over the years. One cannot model without modeling *something*, so when students are engaged in modeling, reasoning processes and scientific concepts are always deployed together.

Most existing research on modeling has been conducted with units or courses that do not span more than one school grade. (For example, Stewart and colleagues have developed high school courses in evolutionary biology and genetics; Reiser et al., 2001; White and Frederikson, 1998; Raghavan et al., 1995; and Wiser, 1995 have developed units for middle school grade students.) On a longer time scale, Lehrer and Schauble (2000) have initiated and studied a school-based program in which science teaching and learning is organized over grades 1-6 around modeling approaches to science (see Box 4.8). Data from this project include paper-and-pencil "booklet" items administered to intact classes of students, yearly three-hour detailed student interviews, and "modeling tasks" completed by small groups of students. Producing these items was itself a challenging task, since students were learning forms of mathematics not routinely taught in elementary grades. The items that were developed were based on evolving data about children's understanding of ideas in geometry, measurement, data, and statistics. The student achievement data showed strong student gains; for example, from the first to the second year of the project, effect sizes by grade were 0.56 (Grade 1), 0.94 (Grade 2), 0.43 (Grade 3), 0.54 (Grade 4), and 0.72 (Grade 5).

Argumentation Bazerman (1988), Lemke (1990), Kuhn (1989), and others have pointed out that science entails mastering and participating in a particular form of argument, including relationships between theories, facts, assertions, and evidence. This characterization of science explicitly acknowledges that science is not just the mastery of knowledge, skills, and reasoning but also participation in a social process that includes values, history, and personal goals. This view of science informs the ongoing work of Warren and Rosebery (1996), for example, who focus on classroom discourse organized around argumentation in science (see Box 4.9). Once again, researchers are supplementing their reports of teachers' professional development with careful measures of student learning. These measures are specific to both the content that students are studying in their classrooms (e.g., interviews of students' grasp of ideas about motion) and in general (e.g., noting changes in the rates of certain patterns of discourse in classroom discussions of science).

• •

BOX 4.8 Science as Modeling

In Lehrer and Schauble's (2000) program, researchers work with teachers to reform instruction and, in coordination, to study the development of model-based reasoning in students. Early emphasis is on developing young children's representational resources (drawings, writing, maps, three-dimensional scale models) as they conduct inquiry about aspects of the world that they find theoretically interesting.

For example, first graders studied ripening and rot by using drawings to record changes in the color and squishiness of fruit, compost columns to investigate rates of decomposition, and maps of the school to investigate the dispersal of fruit flies from the compost columns to classrooms near and far. Teachers typically begin modeling with young students by exploring models that literally resemble the scientific phenomena being modeled. For example, first graders cut green paper strips to record changes over time in the height of amaryllis and paperwhite narcissus that they grew in soil and in water. Then, as investigations proceed, initial models are successively revised to provide increased representational power.

As in the history of modern science, these models increasingly incorporate mathematical descriptions of the world. The students investigated concepts about measurement as they investigated which of the plants grew tallest (Lehrer and Schauble, 2000). However, when the teacher shifted the question to "Which plant grew fastest?" attention turned to recording and representing changes in height over time. In subsequent grades, questions about plants expanded to include comparison of growth rates (with attention to logistic curves as a general model of growth), the volume of their canopies (investigations about whether canopies grow in geometrically similar proportions), and shapes and other qualities of distributions of plants grown under different conditions (including sampling investigations).

Researchers are investigating the potential of a range of central science themes (growth and diversity, animal and human behavior, structure) that can support this kind of cumulative modeling approach. The objective is to develop a cumulative approach to science that permits steady growth in students' modeling repertoires across the elementary and middle school grades. One focus of research is to identify themes that are central to later science instruction and that provide early entry to young students and smooth "lift" (increased challenge) as students graduate from grade to grade. The primary form of professional development in this program is teachers' collective investigation of the development of student thinking and study of the implications of those findings for teaching. The research also tracks the professional development of participating teachers and documents the institutional conditions required to support these forms of teaching and learning.

BOX 4.9 Science as Argumentation

The Cheche Konnen project developed by Warren and Rosebery (1996) turns the attention of practicing teachers toward student meaning-making in science, especially those students whose first language is not English. Instruction capitalizes on students' linguistic and cultural resources developed outside the school. In addition, teachers are encouraged to emphasize that the work of practicing scientists is also "populated by intentions, those of the speaker and those of others, both past and present" (p.101). Teachers seek to find points of contact between their students' talk and reasoning and the forms of communication observed in communities of professional scientists. By conducting their own extended scientific inquiries, teachers in the Cheche Konnen project come to better understand the social and human basis of the scientific enterprise. Together, teachers conduct close study of student language by analyzing and investigating videotapes of classroom discourse. The assumption in this work is that student talk is sensible, and that the teacher's job is to become increasingly skilled at identifying that sense and using it as the foundation for instructional moves.

Checkpoints: Assessment

As in other subjects, quality assessment in science requires, as a starting point, an understanding of what students should know and be able to do. It is perhaps not surprising, then, that the current situation in science assessment outside physics is dire. The broad but shallow coverage of science topics in current texts is mirrored in standardized assessments (including those administered for accountability purposes by the states) that touch briefly on a very wide array of concepts and topics without deeply probing student understanding of any of them. Some assessments include items designed to tap common student misconceptions, but they do not diagnose the developmental level of a student's thinking about the topic.

The diagnosis of student understanding that would render an assessment of greater use for instruction would be difficult to achieve without narrowing the range of topics. In-depth assessment, like that done in the Force Concept Inventory (discussed above) of so many topics, would not be practical in a single assessment. The current practice of devoting no more than a few items to each of several topics means that the assessments do not capture information that teachers can use. Even worse, they may serve to reify bad practice by encouraging an instructional emphasis on coverage over conceptual understanding. There have been recent attempts to develop alternative performance-based assessments in science to address these problems, but these have not yet overcome the psychometric and logistical challenges (Ruiz-Primo and Shavelson, 1996; Solano-Flores and Shavelson, 1997; Ruiz-Primo et al., 2001; Stecher et al., 2000). Moreover, attempts to change the form of assessment are hampered by the more fundamental problem of forging consensus about what is worth assessing.

TEACHER KNOWLEDGE

Relatively few teachers at the K-8 level feel well qualified to teach life, earth, or physical science. The percentages range from 18 for physical science to 29 for life science. K-8 teachers generally lack a deep knowledge of the subject matter of science. Few have an undergraduate major in a science discipline, although most have done some science coursework while in college. Undergraduate courses in science are not particularly helpful for understanding the rich conceptual repertoires that children typically bring to understanding scientific situations.

There is little in teacher preparation programs that provides the foundations of pedagogical content knowledge for teaching science. Elementary school teachers are less likely than middle or high school teachers to indicate that they are prepared to support the development of students' conceptual understanding of science, provide deep coverage of fewer science concepts, or manage a class of students engaged in an extended inquiry project (Weiss et al., 2001). The available evidence does suggest, however, that "teachers who participate in standards-based professional development often report increased preparedness and increased use of standards-based practices, such as taking students' prior conceptions into account when planning and implementing science instruction. However, classroom observations reveal a wide range of quality of implementation among those teachers" (Horizon Research, 2002:168-169).

RESEARCH AGENDA

International and national test scores highlight the weakness of K-12 science education in the United States. That students in so many other countries perform considerably better suggests that the problems are tractable. And there are some indicators of the path toward improvement: we need to be more thoughtful about supporting a deeper understanding of the big ideas in science curricula. This implies principled choice of fewer topics to be treated in greater depth and with greater coherence. These must then be the ultimate targets of a SERP agenda that holds promise for improving student learning outcomes.

Promising work, examples of which are described, has begun to build the knowledge base for a more coherent approach to science education. Teams of researchers working closely with teachers and other educational practitioners are systematically exploring the long-term learning potential and technical feasibility of pursuing systematic, cumulative approaches to science that treat topics in depth. All of these efforts include thoughtful consideration of the appropriate goals of early science education, investigations of the development of student knowledge, and research on the professional development and institutional supports required to implement them. Finally, each of these efforts relies on practitioner-researcher partnerships that extend over a number of years. Such long-term relationships are essential because the targets of the research (forms of student thinking) must first be reliably generated before they can be systematically studied. In these many respects, these are the types of efforts we have argued carry potential to improve practice.

So far, each of these efforts has been patched together with the short-term grant awards that are typical in education funding. They have not had the support required for evaluating long-term student outcomes rigorously and independently, with a broad range of students in a range of settings. But they provide a very promising point of departure for a SERP research program.

With its longer time scale and capability to plan comparative studies of the trade-offs of different approaches, SERP could assume a critical role in the development, study, and comparison of these models of teaching and learning that build systematically across the grades of schooling. Some of the programs we have mentioned are already yielding longitudinal findings about student learning. Some are investigating the forms of professional development and institutional support that are required to help similar programs flourish more widely. For this work to contribute to the quality of K-12 science education on a large scale, however, will require a sustained effort to learn from the range of experiments and to use what is learned to inform both the next stages of research and development and the goals and standards set for science learning. To this end, the research agenda we propose involves initiatives that we spell out in general terms, and that will look in their specifics much like the initiatives in reading and mathematics:

- Development and evaluation of integrated learning-instruction models, with component efforts that include curriculum, assessment, and teacher knowledge requirements;
- Evaluating standards for science achievement.

Initiative 1: Development and Evaluation of Integrated Learning-Instruction Models

Identifying a productive organizing core for school science across the grades is an important element in providing science education that builds from one year to the next. This does not suggest that there is a single, right vision about what is worth teaching and learning. But alternative visions should be formulated, articulated, and carefully justified, so that instruction in all cases can be oriented around valued goals. However, the challenges and possibilities of alternative commitments become clear only when the details of instruction have been worked out, conjectures about fruitful paths for learning have been developed and pursued, and longitudinal research has been conducted as instruction plays out in classrooms. While there have been several efforts to establish standards (the core), these have had minimal impact because they are promulgated without the details of instruction that are required to attain the goals that are envisioned. This will require work on curriculum and on assessment that is closely linked.

Curriculum development and evaluation Existing promising programs like those described above should be further developed and evaluated. The work we propose is not the typical process in which curricula are first invented and then evaluated. Rather, it is one in which design and research are intimately interleaved, so that initial design decisions take the status of conjectures. Evidence regarding the conjectures and their consequences for learning would then contribute to the ongoing shaping and revision of the design. Such an approach is particularly important for a field like science education, in which there is not yet an organized research base to undergird conjectures about optimal sequences of topics and tasks. The key role to be played by SERP is in evaluating the range of programs to consolidate an understanding of differences across programs and their implications for student learning outcomes.

The nature of the work to be done here parallels several of the research and development efforts described in previous chapters. The evaluation of curricula for science would be much like the evaluation of curricula in mathematics. The desired outcome here, as there, is not a stamp of success or failure, but a deeper understanding of the learning process and effective avenues to support it, with a goal of continuous learning and improvement.

The work should be designed to collect outcomes data longitudinally. Different instructional commitments necessarily produce different results, but it is difficult to evaluate those results unless we can see what they generate over the long term. Excellent science instruction achieves more than the development of relevant concepts; it also fosters habits of mind that are consistent with scientific ways of knowing. These forms of thinking are acquired only over years of systematic support and assistance. For this reason, we cannot understand the potential payoff of the varying approaches to science education unless the contexts permit sober estimation of what they deliver over the long term.

Initiative 2: Assessment

An essential limitation on the new experiments in science education is that they lack a widely shared set of assessment instruments (like the Force Concept Inventory in physics) that can anchor meaningful comparisons across different approaches. One potential source for such a set of assessments might be the instruments and items recently designed to assess the development of students' understanding of scientific epistemology. These include interviews about the nature of science and the nature of models and their uses in science, both developed by Carey and her colleagues (Carey et al., 1989; Grosslight et al., 1991) and by Rosalind Driver's research group (Driver et al.,

1996). In each case, these interviews take a long-term developmental focus and have generated baseline data on students in a wide variety of classrooms.

These existing evaluations, while promising for the purposes for which they were designed, focus primarily on the nature of science rather than science content. Some work has been done that could support assessment development on the latter, however. Studies assessing students' conceptual development have reported a variety of tasks and items designed to measure conceptual development within particular subject-matter areas. An example is the interview developed by Vosniadou and Brewer to diagnose children's conceptual models of the earth-moon-sun (Vosniadou and Brewer, 1994). Obviously, the utility of these assessments depends on their specific relevance to the material being taught in the classroom, but they do serve as a potential source of both items and approaches to assessment development.

One potential role for SERP would be to convene researchers and curriculum developers in science who would agree to develop and commit to the use of a common set of assessments. The programs of group members would need to demonstrate at least partial overlap in both conceptual content and epistemological focus. This could be done, for example, with several of the elementary school programs described earlier that emphasize some common approaches to the study of animal and human behavior, adaptation, and evolution and that share some common commitments about the nature of science. The idea would be not to identify a set of assessments that would serve once and for all to measure learning in science, but to serve as a test case for the possibility of developing and refining at least one powerful science assessment that takes a developmental approach to measuring the evolution of student knowledge and understanding. Moreover, such assessments would be fundamentally important in pursuing the study of trade-offs of different commitments to an organizing core for science education.

The SERP network would be a natural site for such research because SERP can support the kind of long-term effort needed to develop and test assessments and can also bring to bear the multiple student data sets and forms of professional expertise (e.g., psychometricians, content experts) required.

Initiative 3: Teacher Knowledge Requirements

As in reading and mathematics, little attention has been given to the knowledge requirements to effectively teach science at any grade level or to effectively connect what is learned at one level with what has come before and what will come after. The avenue that we propose for exploring these questions is to investigate the knowledge requirements to effectively work with the curricula under study in a program of instruction.

The research and development programs discussed above that are currently operating in schools provide a rich resource in the form of teachers who can be studied. Each of these programs has made commitments to forms of professional development, and it would be highly informative to compare the different approaches. For example, some of them work with volunteer teachers from a wide geographical area, for example, across large school districts. Others work with every teacher in a participating school. It would be important to understand the relative advantages of working with a selective, presumably very committed group versus the potential synergies to be gained from working with an intact school staff. A variety of other professional development strategies are being used and studied in these programs, including teacher authoring, science learning workshops, study of student work, reading of articles and texts about science and science education, and analysis of discourse on classroom videotapes. Little serious comparative study has been conducted of the relative costs and benefits of such strategies in spite of the obvious importance of such information for policy makers and administrators.

As in the other domains, the work must provide an evidence base on the knowledge required and the knowledge that is typical of science teachers at different grade levels. The distinction between what teachers themselves know about science and what they know about how to teach science to a student will be as critical, as it is in mathematics. (The descriptions of research on teacher knowledge in the mathematics and reading chapters, as well as the physics section of this chapter, provide more detail regarding the nature of the questions to be examined and the approaches to teacher education that should be compared.)

Initiative 4: Evaluating Standards for Science Achievement

The practitioners and researchers engaged in the study of science instruction and student learning will be well placed to inform the ongoing efforts by various stakeholders to set standards for student achievement in science at various grade levels. The standards themselves are, and should be, based on more than research. Much depends on society's educational goals for its children and the relative importance it places on competing goals. But goal setting can be more rational if it is well informed. One strand of the SERP science work we propose is the consistent attention to, and articulation of, what is possible and with what commitments (investment in teacher education, instructional time, etc.). This should be done through careful data collection and regular stock-taking of results across studies.

5 Ensuring Quality and Impact: A Program to Advance Both Science and Practice

Much has been made in recent years of the quality of education research, with particular emphasis on the methodological weakness that is said to characterize the field. The discussion is a complex one, for issues of quality are regularly confounded with issues of complexity. Education needs high-quality research if the results are to be reliable for purposes of improving practice. Research also needs to be designed to be sensitive to the contexts of teaching and learning if it is to be of use to teachers.

In this chapter we articulate the features of the SERP organization and the proposed agenda that together support research quality and impact.

At the request of the U.S. Department of Education, a committee was recently convened at the National Research Council to examine and clarify the nature of scientific inquiry in education. Its report, *Scientific Research in Education* (National Research Council, 2002b), describes features of what that committee considered high-quality research. The committee's six principles can serve as a point of departure for our discussion (National Research Council, 2002b):

- Pose significant questions that can be investigated empirically;
- Link research to relevant theory;
- Use methods that permit direct investigation of the question;
- Provide a coherent and explicit chain of reasoning;

- Replicate and generalize across studies; and
- Disclose research to encourage professional scrutiny and critique.

The principles are straightforward and are sound elements of scientific research in any discipline. However the challenge of conducting research that is responsive to these principles in education is somewhat daunting. The SERP proposal and research agenda provide the specifics with which to put flesh on these skeletal principles of quality research and consider how quality in education research might be effectively supported.

POSE SIGNIFICANT QUESTIONS THAT CAN BE INVESTIGATED EMPIRICALLY

The significance of the questions we pose is heightened by the consistent focus in the agenda on the questions that define educational practice. A systematic approach to reviewing what is known and unknown regarding those questions in each domain of study draws attention to research questions that are critical from the perspective of improving teaching and learning.

But while significant questions in education research are many, the ability to investigate those questions empirically is more constrained. Understanding effective instructional practice, for example, requires access to classrooms in which that practice can be observed. It requires careful data collection regarding the many features relevant to instruction, as well as the many features of the students themselves that may affect instructional outcomes.

Challenges abound. Putting data collection systems in place requires a substantial up-front investment. That investment may make little sense when a school or school district is participating in an individual study. Only in rare circumstances will the benefit to the researcher and that to the school warrant the cost. However, if a SERP organization and research program is in place, schools and school districts that function as field sites would have an ongoing involvement in the research enterprise. The data collected could be used in many different studies, making the initial investment far more productive.

Moreover, the value of the data collection rises substantially when students are followed over many years. Longitudinal data sets are the workhorses of empirical research in the social sciences. Investigation of the long-term impacts that are central to effective policy making can be explored only when data are collected longitudinally. But because longitudinal data collection requires a long-term effort in an environment in which many of the individuals (researchers, principles, teachers) have short-term horizons, the presence of an institutional infrastructure that can ensure continuity is critical.

Still, schools may be reluctant to participate in data collection without the promise of both protection (privacy of information) and significant payoff. Access to the quality of data needed for direct investigation of instructional questions becomes more likely if an organization like SERP develops solid credentials at instituting and honoring standards for information privacy—even as it attends to making data maximally useful for researchers within those privacy constraints.

Payoff for schools will be required in the form of negotiated products of research and development that satisfy the needs of schools as well as those of the researchers. It is certainly likely that many schools would have no interest in a research partnership. But the school administrators who do look to the research community for help with puzzling questions of practice have little guidance about where to find that help. SERP can serve as a magnet for schools and school districts who are looking for partnerships with researchers. When the desire for partnership is mutual, the demands of careful data collection may be more easily met.

A major contributor to both quality and impact of the SERP program, then, is the ability of the organization to attract schools as field sites and to nurture long-term relationships. Cultivating those relationships will be critical in achieving the trust that will be required for access to classrooms and to student-level data that will make empirical investigation of the most important problems of practice possible.

LINK RESEARCH TO RELEVANT THEORY

The National Research Council report (2002b) argues, "it is the long-term goal of much of science to generate theories that can offer stable explanations for phenomena that generalize beyond the particular." In the practice of education, much changes from one case to the next: students, teacher, principal, curriculum, textbooks, school environment, and home environment can all vary simultaneously. Progress in advancing theory can be challenging because finding stable relationships can be difficult when so little is held constant. Yet the ability to generalize beyond the particular is precisely what is needed not only for the quality of the research, but for supporting teaching as a profession.

Several features of the proposed agenda support a close link between research and theory. First, in the three domains of focus, we describe knowledge bases that provide a theoretical and empirical foundation relevant to each of the framing questions. In each domain the proposed research and development builds on that foundation, and on subject specific theory and knowledge. In some cases (e.g., the acquisition of early number knowledge) theory is well developed and empirically supported, and we suggest that curricula *built on the theoretical underpinnings* are ripe for further work. In other areas (e.g., reading comprehension), the theory is in dispute. The research proposed (e.g., observations of effective practice) can support theory building. In all cases, however, the starting point for the proposed work is a review of the current state of theory and knowledge, ensuring that the link between the research and relevant theory is strong.

Second, development and evaluation of instructional interventions can be, and often are, carried out with the sole purpose of discovering "what works." Advancing theory, however, requires knowledge of *why* something works, for whom, and under what circumstances. The agenda we propose systematically pursues that theory-building knowledge.

Finally, theory development will be supported by the rich body of research that will be conducted in networks that are closely linked. The work conducted separately for science, mathematics, and reading can provide a rich knowledge base from which new understandings and generalizations can emerge, moving the theoretical understanding of teaching and learning forward. The discussion of issues in early reading, for example, in many respects mirrors that in early mathematics. Both subjects pinpoint the need to develop fluency in a skill and yet not hold more challenging thinking about problems or text hostage to skill development. Both struggle with the problem of excessive focus on procedure and insufficient attention to meaning-making. The challenge for teachers to engage students in dialogue that takes their thinking the next step is a close parallel in these two, and other, subject areas. Patterns across subjects can support new theoretical understandings. The responsibility of the leadership of the networks for monitoring and synthesizing findings across the many strands of work will allow SERP to advance theory development more intentionally than is now the case.

USE METHODS THAT PERMIT DIRECT INVESTIGATION OF THE QUESTION

The Committee on Scientific Principles for Education Research argues, quite sensibly, that there is no best method of research in education or in any other field. Rather, the method must be appropriately matched to the research question asked. In the agenda outlined here, the proposed methods vary tremendously in accordance with the question. In some cases in which the knowledge base is weak, we suggest observation of practice to support hypothesis generation. In other areas in which curricula have been developed and subjected to preliminary evaluation, we suggest randomized trials in order to determine effectiveness of a curriculum.

Perhaps the more important point made by the report on scientific research in education is that rarely can a single method illuminate all the questions and issues in a line of inquiry. Indeed, our agenda proposes that a range of questions and companion methods are required as components of a *program* of research and development in order for research and practice to be powerfully linked. While qualitative research intended, for example, to generate hypotheses regarding effective teaching practice may be desirable when theory and evidence are weak, the product of that research can be broadly useful to practice only when the hypotheses are rigorously tested. Similarly, a randomized trial to determine the learning outcomes of alternative curricula can provide high-quality data, but fully exploiting the findings and generating further improvements in the curriculum and in teaching practice requires observation of why and how teaching and learning change with the curriculum. Quality and impact adhere in the program that combines methods to answer the different kinds of questions critical to improving practice.

PROVIDE A COHERENT AND EXPLICIT CHAIN OF REASONING

The explanations and conclusions drawn from research, the NRC committee points out, "requires a logical chain of reasoning from evidence to theory and back again that is coherent, shareable, and persuasive to the skeptical reader" (p. 4). The strength of that chain of reasoning is critical in education research, because beliefs are often held passionately, and any case made for improving the teaching of mathematics, reading, or science will encounter a great many skeptics. Ultimately, however, many of the contested claims can be tested empirically. But because skepticism runs high, high-quality evidence will be required for impact.

Central to the quality of the evidence in the research we propose here is an investment in the development of credible outcome measures and control variables. If we want students to have a deeper understanding of science, a more flexible approach to mathematical problem solving, or an appreciation of nuance and complexity in text, we must have adequate measures of those outcomes to know if they have improved. Similarly, if we want to understand the conditions under which an instructional approach improves outcomes, we need to have carefully defined which "conditions" matter. These investments are critical to the quality of the research, and yet they are often shortchanged in poorly funded research studies.[1]

---

[1]See, for example, the discussion of evaluations of curricula funded by the National Science Foundation in Chapter 3.

The quality of the evidence also improves substantially when data are collected longitudinally. As noted earlier, longitudinal data are required for measuring long-term results. But they can also be useful in suggesting patterns that might otherwise go unnoticed, calling attention, for example, to years or classrooms that are outliers and so hold information that may be valuable. And since much of what can be achieved in later years depends on what has been taught in earlier years, the ability to look across years is fundamental to designing instruction that effectively builds over time. The systematic, longitudinal data collection effort that SERP proposes can bolster the quality of the evidence from its research.

But even findings from good-quality data require interpretation and inference. The nature of the SERP networks as collaborations among researchers and practitioners will provide a fertile environment for interpretation of findings that draws on different perspectives, knowledge bases, and experiences.

REPLICATE AND GENERALIZE ACROSS STUDIES

The Committee on Scientific Principles for Education Research argued that ultimately, scientific knowledge advances when findings are reproduced in a range of times and places and when findings are integrated and synthesized (National Research Council, 2002b). While replication is required for scientific quality, it is at the very heart of impact on practice. As we have argued throughout this report, to know that an instructional program works well for students in a middle-class, suburban school provides little useful knowledge to teachers whose limited-English-proficient students live in a disadvantaged urban environment. A hallmark of the agenda we propose here is that it consistently builds replication of research results into each stage of the work. Moreover, we propose that the research be conducted with the range of students and under the range of circumstances to which the results would apply.

Both replication and generalization are facilitated by the SERP organization. An important function of the networks is to develop common research protocols to facilitate replication and strengthen interpretability across sites and studies. And a central function of the network leadership as described in the SERP report, *Strategic Education Research Partnership,* includes regular synthesizing of research results and stock-taking to alter or chart a course.

· ·

DISCLOSE RESEARCH TO ENCOURAGE PROFESSIONAL SCRUTINY AND CRITIQUE

Few would argue that this widely embraced norm of scientific research is critical to quality. And indeed, the SERP report proposes a scientific advisory board with responsibilities for quality monitoring and professional scrutiny of the research program. In addition to a formal review process, the SERP design includes a proposed web site that will provide an opportunity for feedback from a wide audience—skeptics included.

The scrutiny that is required for impact in the field of education, however, goes beyond the disclosure of research results that quality mandates. Much of the education research produced, both high-quality and otherwise, withers from neglect. Impact requires actively seeking out high-quality research that is important for educational practice and building on it—as we propose, for example, with respect to the Number Worlds research or the reciprocal teaching research.

But impact also requires the design of research studies that can take the knowledge from practice and incorporate it into testable propositions that can be shared publicly. Much of what teachers learn from repeated observation of student learning and response to instruction is never formally articulated, tested, or shared with others in their professional community (Hiebert et al., 2002). Research designed to learn from practice, like that proposed for reading comprehension, can formalize the knowledge of teachers, subject it to testing, and make those results available to be shared and scrutinized publicly.

Ultimately, both quality and impact are best supported not in single research studies, but in a coherent program of research that ensures replication, accumulation, and follow-through; a program with strong theoretical underpinnings. The quality of the program is elevated substantially when careful attention is paid to research protocols, the design and testing of outcome and control variables, and careful data collection. But at the heart of quality and impact on education practice is an investment in both problems that matter to practice and in the development of communities of researchers and practitioners to carry the work forward. It is this that the SERP design calls for, and this that our proposed agenda will require.

· ·

CONCLUSION

As we consider what we know about high-quality educational practice in reading, mathematics, and science, the gaps in that knowledge base are striking. Widespread concern about the performance of the education system has led states and school systems to develop content and performance standards, and it has led to efforts at both the federal and state levels to hold schools accountable for results. But setting goals is unlikely to be helpful in the absence of a defined path to achieving them.

We have seen repeatedly in the chapters on reading, mathematics, and science that much path-defining work remains to be done. In all of the subject areas, for example, aspects of instruction that are fairly easily defined claim a disproportionate share of instructional time and dominate assessments. These tend to be procedural, and they can therefore be straightforwardly described. Teaching phonics, reproducing factual detail of a story, and teaching math facts or science facts are all cases in point. These are critical dimensions of instruction, but if they are the whole of instruction, then understanding will be shallow (National Research Council, 2000).

This instructional emphasis on the procedural has persisted over the history of public schooling, although rising standards suggest the ultimate goal of instruction is to foster deeper understanding and skill: to see nuance in a text's meaning, to appreciate the difference between a finding and an opinion, to understand the nature of mathematical or scientific problems and how they are identified. But building understanding is, without question, challenging. It will require more interaction between teacher and students, as programs like questioning the author, reciprocal teaching, and modeling methods suggest. It will require making student thinking apparent, and working with student ideas to take them to the next level of sophistication, as Everyday Mathematics or ThinkerTools suggest. Some teachers have the intuition, inclination, knowledge, and experience to provide their students with these types of instructional experiences. But for the many teachers who do not, the supports that would allow them to develop these abilities are rarely available. Yet success at reaching high academic standards depends on doing just that.

On some of the most basic instructional questions—for example, how can the components of reading comprehension be developed and assessed across the school years?—we have hardly begun the research necessary to support professional practice. Of equal concern, however, is that in those areas in which the knowledge base for improving practice is strong—like early mathematics, early reading, and physics—it has had little impact on practice. The examples we highlight of high-quality research and development have, for the most part, remained on the sidelines of educational practice. Perhaps the greatest squandered resource, however, is the excellent teaching practice that produces demonstrable effects on student achievement and yet remains an anomaly, its lessons for teaching never articulated, studied, or shared.

The work we propose here has the potential to substantially improve the knowledge base to support teaching and learning by pursuing answers to questions at the core of teaching practice. It calls for the linking of research and development—of instructional programs, assessment tools, teacher education programs and materials—that is now rare. In the course of doing so, we propose to draw on the largely untapped resources of effective teaching practice and high-quality research. The downstream proposals offer hope of an impact in the relatively near term, for they build on solid work with clear implications for practice. And the upstream proposals would begin to provide a foundation for professional practice in the future where none now exists.

The panel generated this agenda assuming the infrastructure and operation of a SERP organization to support the linked, multifaceted work that is envisioned. Realizing the potential of the proposed research and development for improving teaching and learning will require the organizational infrastructure that can support research on, with, and for practice.

References

Adams, J., R. Treiman, and M. Pressley
   1998   Reading, writing, and literacy. Pp. 275-355 in *Handbook of Child Psychology, Fifth Edition, Vol. 4: Child Psychology in Practice,* I.E. Sigel and K.A. Renninger, eds. New York: Wiley.

Aleven, V., and K. Koedinger
   2002   An effective metacognitive strategy: Learning by doing and explaining with a computer-based cognitive tutor. *Cognitive Science* 26:147-179.

American Association for the Advancement of Science
   1991   *Science for All Americans,* F.J. Rutherford and A. Ahlgren, eds. Cary, NC: Oxford University Press.
   2001   *Designs for Science Literacy.* Cary, NC: Oxford University Press.

Anderson, J.
   1983   *The Architecture of Cognition.* Cambridge, MA: Harvard University Press.

Anderson, J.R., A. Corbett, K.R. Koedinger, and R. Pelletier
   1995   Cognitive tutors: Lessons learned. *Journal of the Learning Sciences* 4:167-207.

Anderson, J.R., L.M. Reder, and H. Simon
   1998   Radical constructivism and cognitive psychology. In *Brookings Papers on Education Policy 1998,* D. Ravitch, ed. Washington, DC: Brookings Institute Press.

Ashby, R., P. Lee, and D. Shemilt
   Forth-   Learning history: Principles into practice. Chapter in *How Students*
   coming   *Learn: History, Math, and Science in the Classroom.* Committee on How People Learn, A Targeted Report for Teachers. Washington, DC: National Academies Press.

Avorn, J., and D.H. Solomon
   2000   Cultural and economic factors that (mis)shape antibiotic use: The nonpharmacologic basis of therapeutics. *Annals of Internal Medicine* 133(2):128-135.

Avorn, J., and S.B. Soumerai
   1983   Improving drug-therapy decisions through educational outreach: A randomized controlled trial of academically based "detailing." *New England Journal of Medicine* 308:1457-1463.

Avorn, J., M. Chen, and R. Hartley
   1982   Scientific versus commercial sources of influence on the prescribing behavior of physicians. *The American Journal of Medicine* 73:4-8.

Ball, D.L.
 1988 *The Subject Matter Preparation of Prospective Mathematics Teachers: Challenging the Myths*. Research Report 88-3. East Lansing: Michigan State University, National Center for Research on Teacher Learning. Available: http://ncrtl.msu.edu/http/rreports/html/rr883.htm.

Barr, M.A., S. Ellis, H. Tester, and A. Thomas
 1988 *The Primary Language Record: Handbook for Teachers*. Portsmouth, NH: Heinemann.

Bazerman, C.
 1988 *Shaping Written Knowledge: The Genre and Activity of Experimental Article in Science*. Madison: University of Wisconsin Press.

Beck, I.L., and M.G. McKeown
 2001 Text talk: Capturing the benefits of read-aloud experiences for young children. *The Reading Teacher* 55(1):10-20.

Beck, I.L., M.G. McKeown, R.L. Hamilton, and L. Kucan
 1997 *Questioning the Author: An Approach for Enhancing Student Engagement with Text*. Newark, DE: International Reading Association.

Bednarz, N., C. Kieran, and L. Lee
 1996 *Approaches to Algebra: Perspectives for Research and Teaching*. Dordrecht, The Netherlands: Kluwer Academic Press.

Beichner, R.J.
 1994 Multimedia editing to promote science learning. *Journal of Computers in Mathematics and Science Teaching* 13(2):147-162.

Bishop, D.V.M., and C. Adams
 1990 A prospective study of the relationship between specific language impairment, phonological disorders and reading retardation. *Journal of Child Psychology and Psychiatry* 31:1027-1050.

Boaler, J.
 1997 *Experiencing School Mathematics: Teaching Styles, Sex, and Setting*. Buckingham, UK: Open University Press.

Bos, C.S., N. Mather, R.F. Narr, and N. Babur
 1999 Interactive, collaborative professional development in early literacy instruction: Supporting the balancing act. *Learning Disabilities Research and Practice* 14(4):227-238.

Bowey, J.A.
 1994 Phonological sensitivity in novice readers and nonreaders. *Journal of Experimental Child Psychology* 58:134-159.

Brenner, M.E., R.E. Mayer, B. Moseley, T. Brar, R. Duran, B.S. Reed, and D. Webb
 1997 Learning by understanding: The role of multiple representations in learning algebra. *American Educational Research Journal* 34(4):663-689.

Briars, D.J., and L.B. Resnick
 2000 *Standards, Assessment—and What Else? The Essential Elements of Standards-Based School Improvement*. Los Angeles: Center for the Study of Evaluation, University of California, Los Angeles. Available: http://www.cse.ucla.edu/CRESST/Reports/TECH528.pdf.

Brooks-Gunn, J., G.J. Duncan, and P.R. Britto
 1999 Are socioeconomic gradients for children similar to those for adults? Pp. 94-124 in *Developmental Health and the Wealth of the Nations: Social, Biological, and Educational Dynamics*, D.P. Keating and C. Hertzman, eds. New York: Guilford Press.

Brown, A.L.
1997 Transforming schools into communities of thinking and learning about serious matters. *American Psychologist* 52:399-414.

Brown, A.L., J.D. Bransford, R.A. Ferrara, and J.C. Campione
1983 Learning, remembering, and understanding. Pp. 78-166 in *Handbook of Child Psychology: Volume 3 Cognitive Development* (4th edition), J.H. Flavell and E.M. Markman, eds. New York: Wiley.

Budiansky, S.
2001 The trouble with textbooks. *Prism* (February).

Carey, S.
2001 Evolutionary and ontogenetic foundations of arithmetic. *Mind and Language* 37-55.

Carey, S., R. Evans, M. Honda, E. Jay, and C. Unger
1989 An experiment is when you try it and see if it works: A study of grade 7 students' understanding of the construction of scientific knowledge. *International Journal of Science Education* 11:514-529.

Carpenter, T., and M. Franke
2001 Developing algebraic reasoning in the elementary school. Pp. 155-162 in *Proceedings of the 12th ICMI Study Conference* (Vol. 1), H. Chick, K. Stacey, J. Vincent, and J. Vincent, eds. Melbourne, Australia: The University of Melbourne.

Carpenter, T., E. Fennema, and M. Franke
1996 Cognitively guided instruction: A knowledge base for reform in primary mathematics instruction. *Elementary School Journal* 97(1):3-20.

Carroll, G.R., and M.T. Hannan
2000 *The Demography of Corporations and Industries.* Princeton, NJ: Princeton University Press.

Case, R., and R. Sandieson
1987 General Development Constraints on the Acquisition of Special Procedures (and Vice Versa). Paper presented at the annual meeting of the American Educational Research Association, Baltimore, April.

Chi, M.T.H., and M. Bassok
1989 Learning from examples via self-explanations. In *Knowing, Learning, and Instruction: Essays in Honor of Robert Glaser*, L.B. Resnick, ed. Hillsdale, NJ: Lawrence Erlbaum Associates.

Chi, M.T.H., P.J. Feltovich, and R. Glaser
1981 Categorization and representation of physics problems by experts and novices. *Cognitive Science* 5:121-152.

Clay, M.M.
1993 *Reading Recovery: A Guidebook for Teachers in Training.* Portsmouth, NH: Heinemann.

Clement, J.
1982 Students' perceptions in elementary mechanics. *American Journal of Physics* 50:66-71.

Cobb, P., K. McClain, and K. Gravemeijer
2003 Learning about statistical covariation. *Cognition and Instruction* 21(1):1-78.

Cohen, D.K., and H.C. Hill
2000 Instructional policy and classroom performance: The mathematics reform in California. *Teachers College Record* 102(2):294-343.

Cook-Gumperz, J.
1973   Situated instructions: Language socialization of school aged chil-
        dren. Pp. 103-124 in *Child Discourse*, S. Ervin-Tripp and C. Mitchell-
        Kernan, eds. New York: Academic Press.
Cummings, K., D. Halliday, R. Resnick, and J. Walker
2001   *Cummings, Laws, Redish Cooney, Understanding Physics, Part 2 Prelimi-
        nary Edition*. Indianapolis, IN: John Wiley and Sons.
Dauite, C.
1986   Do 1 and 1 make 2? Patterns of influence by collaborative authors.
        Written Communication 3:382-408.
Dickinson, D.K., and K.E. Sprague
2001   The nature and impact of early childhood care environments on the
        language and early literacy development of children from low-in-
        come families. Pp. 263-280 in *Handbook of Early Literacy Research*, S.B.
        Neuman and D.K. Dickinson, eds. New York: Guilford Press.
Dickinson, D.K., and P.O. Tabors
2001   *Beginning Literacy with Language*. Baltimore: Brookes Publishing Co.
diSessa, A.
1982   Unlearning Aristotelian physics: A study of knowledge-based learn-
        ing. *Cognitive Science* 6(1):37-75.
Driver, R., J. Leach, R. Millar, and P. Scott
1996   *Young People's Images of Science*. Buckingham, UK: Open University
        Press.
Durkin, D.
1978–   What classroom observations reveal about reading comprehension
1979   instruction. *Reading Research Quarterly* 14(4):481-533.
Education Development Center, Inc.
2001   *Curriculum Summaries*. Newton, MA: K-12 Mathematics Curriculum
        Center.
Education Week
2002   The new-look SAT. *Education Week on the Web* 21(48)28. Available:
        http://www.edweek.org.
Ehri, L.C.
1998   Grapheme-phoneme knowledge is essential for learning to read words
        in English. Pp. 3-40 in *Word Recognition in Beginning Literacy*, J.L.
        Metsala and L.C. Ehri, eds. Mahwah, NJ: Erlbaum.
Ferguson, R.
1991   Paying for public education: New evidence on how and why money
        matters. *Harvard Journal on Legislation* 28(Summer):465-498.
Ferguson, R.F., and H.F. Ladd
1996   How and why money matters: An analysis of Alabama schools. Pp.
        265-298 in *Holding Schools Accountable*, H. Ladd, ed. Washington, DC:
        Brookings Institution.
Ferrini-Mundy, J., and G. Burrill
2002   Preparing for the Teaching of Algebra in Secondary Schools: Chal-
        lenges and Promising Directions. Workshop report from the Na-
        tional Summit on the Mathematical Education of Teachers: Meeting
        the Demand for High Quality Mathematics Education in America. A
        conference sponsored by the Conference Board of the Mathematical
        Sciences. Falls Church, VA, November 2001. Available: http:
        //www.cbmsweb.org/NationalSummit/WG_Speakers/ferrini-
        mundy.htm.

Fletcher, J.M., B.R. Foorman, A. Boudousquie, M. Barnes, C. Schatschneider, and D.J. Francis
2002   Assessment of reading and learning disabilities: A research-based, treatment-oriented approach. *Journal of School Psychology* 40:27-63.

Foorman, B.R., and L.C. Moats
in press   Conditions for sustaining research-based practices in early reading instruction. *Remedial and Special Education.*

Foorman, B.R., and C. Schatschneider
in press   Measuring teaching practice during reading/language arts instruction and its relation to student achievement. In *Reading in the Classroom: Systems for Observing Teaching and Learning,* S. Vaughn and K. Briggs, eds. Baltimore: Brookes Publishing Co.

Foorman, B.R., and J.K. Torgesen
2001   Critical elements of classroom and small-group instruction promote reading success in all children. *Disabilities Research and Practice* 16:202-211.

Foorman, B.R., J.M. Fletcher, and D.J. Francis
2001   Early reading assessment. In *Testing America's Schoolchildren,* W. Evert, ed. Stanford, CA: The Hoover Institution.

Foorman, B.R., D.J. Francis, K. Davidson, M. Harm, and J. Griffin
2002   Variability in Text Feature in Six Grade 1 Basal Reading Programs. Unpublished manuscript.

Foorman, B.R., J. Anthony, L. Seals, and A. Mouzaki
in press   Language development and emergent literacy in preschool. To appear in *Language Development and Disorders in Childhood,* I. Butler, ed. Special issue of *Seminars in Pediatric Neurology.*

Fuson, K., W.M. Carroll, and J.V. Drueck
2000   Achievement results for second and third graders using the standards-based curriculum "Everyday Mathematics." *Journal for Research in Mathematics Education* 31(3):277-295.

Fuson, K.C.
2003   Children's Math Worlds, Kindergarten, Grades 1, 2, 3, 4, 5. For more information, contact Karen Fuson at Fuson@northwestern.edu.

Gagne, R.M.
1968   Contributions of learning to human development. *Psychological Review* 75:177-191.

Gelman, R.
1967   Conservation acquisition: A problem of learning to attend to the relevant attributes. *Journal of Experimental Child Psychology* 7:167-187.
1990   First principles organize attention to and learning about relevant data: Number and the animate-inanimate distinction as examples. *Cognitive Science* 14:79-106.

Gelman, R., and C.R. Gallistel
1978   *The Children's Understanding of Number.* Cambridge, MA: Harvard University Press.

Ginsburg, H.P., C. Greenes, and Robert Balfanz
2003   *Big Math for Little Kids.* Parsippany, NJ: Dale Seymour Publications.

Gluck, K.
1999   Eye Movements and Algebra Cognitive Tutoring. Doctoral dissertation, Psychology Department, Carnegie Melon University.

Graesser, A.C., K.K. Millis, and R.A. Zwaan
1997   Discourse comprehension. *Annual Review of Psychology* 48:163-189.

Graves, M.F., and W.H. Slater

1987 Development of Reading Vocabularies in Rural Disadvantaged Students, Intercity Disadvantaged Students and Middle Class Suburban Students. Paper presented at the conference of the American Educational Research Association, Washington, DC, April.

Griffin, S.

Forth- Teaching mathematics in the primary grades: Fostering the develop-
coming ment of whole number sense. Chapter in *How Students Learn: History, Math, and Science in the Classroom*. Committee on How People Learn, A Targeted Report for Teachers. Washington, DC: National Academies Press.

Griffin, S., and R. Case

1997 Re-thinking the primary school math curriculum: An approach based on cognitive science. *Issues in Education* 3(1):1-49.

Grosslight, L., C. Unger, E. Jay, and C. Smith

1991 Understanding models and their use in science: Conceptions of middle and high school students and experts. *Journal of Research in Science Teaching* 28:799-822.

Guthrie, J.T., K.E. Cox, E. Anderson, K. Harris, S. Mazzoni, and L. Rach

1998a Principles of integrated instruction for engagement in reading. *Educational Psychology Review* 10(2):177-199.

Guthrie, J.T., P. Van Meter, G.R. Hancock, S. Alao, E. Anderson, and A. McCann

1998b Does concept oriented reading instruction increase strategy use and conceptual learning from text? *Journal of Educational Psychology* 90(2):261-278.

Hake, R.R.

1998 Interactive-engagement versus traditional methods: A six-thousand student survey of mechanics test data for introductory physics courses. *American Journal of Physics* 66:64-74.

Halloun, I.A., and D. Hestenes

1985 The initial knowledge state of college physics students. *American Journal of Physics* 53:1043-1055.

Hannon, B., and M. Daneman

2001 A new tool for measuring and understanding individual differences in the component processes of reading comprehension. *Journal of Educational Psychology* 93:103-128.

Hart, B., and T.R. Risley

1995 *Meaningful Differences in the Everyday Experience of Young American Children*. Baltimore: Paul H. Brookes Publishing Co.

Haverty, L.A.

1999 The Importance of Basic Number Knowledge to Advanced Mathematical Problem Solving. Doctoral dissertation, Psychology Department, Carnegie Melon University.

Heath, S.

1983 *Way With Words*. Cambridge, England: Cambridge University Press.

Heffernan, N., and K.R. Koedinger

1997 The composition effect in symbolizing: The role of symbol production vs. text comprehension. Pp. 307-312 in *Proceedings of the Nineteenth Annual Conference of the Cognitive Science Society*, M.G. Shafto and P Langley, eds. Hillsdale, NJ: Erlbaum.

1998   A developmental model for algebra symbolization: The results of a difficulty factors assessment. Pp. 484-489 in *Proceedings of the Twentieth Annual Conference of the Cognitive Science Society*, M.A. Gernbascher and S.J. Derry, eds. Mahwah, NJ: Erlbaum.

Heibert, E.H., and L.A. Martin
2001   The texts of beginning reading instruction. Pp. 361-376 in *Handbook of Early Literacy Research*, S.B. Neuman and D.K. Dickinson, eds. New York: Guilford Press.

Hestenes, D.
1987   Toward a modeling theory of physics instruction. *American Journal of Physics* 60:440-454.
2000   Findings of the Modeling Workshop Project (1994-00). Final report submitted to the National Science Foundation for the Teacher Enhancement grant entitled Modeling Instruction in High School Physics. Arizona State University, Tempe, AZ.

Hestenes, D., M. Wells, and G. Swackhamer
1992   Force Concept Inventory. *The Physics Teacher* 30:141-158.

Hiebert, J., R. Gallimore, and J.W. Stigler
2002   A knowledge base for the teaching profession: What would it look like and how can we get one? *Educational Researcher* 31(5):3-15.

Horizon Research, Inc.
2002   Unpublished special tabulations on data from the 2000 National Survey of Science and Mathematics Education. Horizon Research, Inc., Chapel Hill, NC.

Hunt, E., and J. Minstrell
1994   A cognitive approach to the teaching of physics. Pp. 51-74 in *Classroom Lessons: Integrating Cognitive Theory and Classroom Practice*, K. McGilly, ed. Cambridge, MA: MIT Press.

Huttenlocher, J.
1998   Language input and language growth. *Preventive Medicine* 27:195-199.

Kalchman, M., and K.R. Koedinger
Forth-   Teaching and learning functions. In *How Students Learn: Targeted*
coming   *Report for Teachers*. Committee on How People Learn, National Research Council. Washington, DC: The National Academies Press.

Kalchman, M., J. Moss, and R. Case
2001   Psychological models for development of mathematical understanding: Rational numbers and functions. Pp. 1-38 in *Cognition and Instruction: Twenty-Five Years of Progress*, S. Carver and D. Klahr, eds. Mahwah, NJ: Erlbaum.

Kintsch, W.
1998   *Paradigms of Comprehension*. Oxford, England: Oxford University Press.

Kintsch, W., K. Rawson, and E. Mulligan
2001   Designs for a Comprehension Test. Paper presented at the Conference of the American Educational Research Association.

Koedinger, K.R., and J.R. Anderson
1998   Illustrating principled design: The early evolution of a cognitive tutor for algebra symbolization. *Interactive Learning Environments* 5:161-180.

Koedinger, K.R., and M.J. Nathan
in press   The real story behind story problems: Effects of representations on quantitative reasoning. *Journal of Learning Sciences*.

Kuhn, T.S.
1989   Possible worlds in the history of science. Pp. 9-32 in Possible Worlds in Humanities, Arts, and Sciences: Proceedings of the Nobel Symposium 65, A. Sture, ed. Berlin: Walter de Gruyter.

Lanier, J., and J. Little
1986   Research on teacher education. Pp. 527-569 in *Handbook of Research on Teaching* (3rd edition), M.C. Wittrock, ed. New York: Macmillan.

Larkin, J.H.
1983   The role of problem representation in physics. In *Mental Models*, D. Genter and A.L. Stevens, eds. Hillsdale, NJ: Lawrence Erlbaum Associates.

Lederman, N.G.
2001   Understanding the Nature of Science: The Potential Contributions to Censorship by Different Views of Science. Paper presented at the International Meeting of the Association for the Education of Teachers in Science, Costa Mesa, CA.

Lehrer, R., and L. Schauble
2000   Developing model-based reasoning in mathematics and science. *Journal of Applied Developmental Psychology* 21(1):218-232.
2001   *Investigating Real Data in the Classroom: Expanding Children's Understanding of Math and Science.* New York: Teachers College Press.
2002   Symbolic communication in mathematics and science: Co-constituting inscription and thought. Pp. 167-192 in *Language, Literacy, and Cognitive Development: The Development and Consequences of Symbolic Communication*, J. Byrnes, and E. Amsel, eds. Mahwah, NJ: Lawrence Erlbaum Associates.

Lemke, J.L.
1990   *Talking Science: Language, Learning, and Values.* Norwood, NJ: Ablex Publishing.

Levidow, B.B., E. Hunt, and C. McKee
1991   The Diagnoser: A hypercard tool for building theoretically based tutorials. *Behavior Research Methods, Instruments, and Computers* 23:249-252.

Lochhead, J., and J. Mestre
1988   From words to algebra: Mending misconceptions. Pp. 127-135 in *The Ideas of Algebra K-12*, A. Coxford and A. Schulte, eds. Reston, VA: National Council of Teachers of Mathematics.

Lonigan, C.J., and G.J. Whitehurst
1998   Relative efficacy of parent and teacher involvement in a shared-reading intervention for preschool children from low-income backgrounds. *Early Childhood Research Quarterly* 13(2):263-292.

Lonigan, C.J., S.R. Burgess, J.L. Anthony, and T.A. Barker
1998   Development of phonological sensitivity in two- to five-year-old children. *Journal of Educational Psychology* 90:294-311.

Lonigan, C.J., S.R. Burgess, and J.L Anthony
2000   Development of emergent literacy and early reading skills in preschool children: Evidence from a latent-variable longitudinal study. *Developmental Psychology* 36:596-613.

Lubienski, C.A., A.D. Otto, B.S. Rich, and P.A. Jaberg
1998   An analysis of two novice K-8 teachers using a model of teaching in context. Pp. 704-709 in *Proceedings of the Twentieth Annual Meeting of the North American Chapter of the International Group for the Psychology of Mathematics Education* (Vol. 2), S. Berenson, K. Dawkins, M. Blanton,

W. Coulombe, J. Kolb, K. Norwood, and L. Stiff, eds. ERIC Document Reproduction Service No. ED 430 776. Columbus, OH: ERIC Clearinghouse for Science, Mathematics, and Environmental Education.

Ma, Liping
   1999   *Knowing and Teaching Elementary Mathematics.* Mahway, NJ: Lawrence Erlbaum Associates.

Maloney, D.P.
   1993   Research on problem solving: Physics. In *Handbook of Research on Science Teaching and Learning,* D. Gabel, ed. New York: Macmillan.

Maloney, D.P., T.L. Okuma, C.J. Hieggelke, and A. Van Heuvelen
   2001   Surveying student's conceptual knowledge of electricity and magnetism. *Physics Education Research, American Journal of Physics Supplement* 69(7):S12-S23.

Mannes, S.M., and W. Kintsch
   1987   Knowledge organization and text organization. *Cognition and Instruction* 4:91-115.

Markman, E.M.
   1977   Realizing you don't understand: A preliminary investigation. *Child Development* 46:986-992.
   1981   Comprehension monitoring. Pp. 61-84 in *Children's Oral Communication Skills,* W.P. Dickson, ed. New York: Academic Press.

Mazur, E.
   1997   *Peer Instruction: A User's Manual.* Edgewood Cliffs, NJ: Prentice-Hall.

McCutchen, D., R.D. Abbott, L.B. Green, N. Beretvas, S. Cox, N.S. Potter, T. Quiroga, and A.L. Gray
   2002   Beginning literacy: Links among teacher knowledge, teacher practice, and student learning. *Journal of Learning Disabilities* 35(1):69-86.

McDermott, L.C.
   1984   Research on conceptual understanding in mechanics. *Physics Today* 37:24-32.

McDermott, L.C., and E.F. Redish
   1999   RL-PER1: Resource letter on physics education research. *American Journal of Physics* 67(9):755-767.

McGlynn, E.A., S.M. Asch, J. Adams, J. Keesey, J. Hicks, A. De Cristofaro, and E.A. Kerr
   2003   The quality of health care delivered to adults in the United States. *The New England Journal of Medicine* 348(26):2635-2645.

Meisels, S.J.
   1996–   Using work sampling in authentic assessments. *Educational Leader-*
   1997   *ship* (Winter):60-65.

Metz, K.E.
   2000   Young children's inquiry in biology: Building the knowledge bases to empower independent inquiry. Pp. 371-404 in *Inquiry Into Inquiry Learning and Teaching in Science,* J. Minstrell and E.H. vanZee, eds. Washington, DC: American Association for the Advancement of Science.

Minstrell, J.
   1989   Teaching science for understanding. Pp. 129-149 in *Toward the Thinking Curriculum: Current Cognitive Research. 1989 Yearbook of the Association for Supervision and Curriculum Development,* L. Resnick and L. Klopfer, eds. Washington, DC: Association for Supervision and Curriculum Development.

1992   Facets of students' knowledge and relevant instruction. Pp. 110-128 in *Proceedings of the International Workshop on Research in Physics Education: Theoretical Issues and Empirical Studies*, R. Duit, F. Goldberg, and H. Niedderer, eds. Kiel, Germany: Institut fur die Padagogik der Naturwissenshaften.

Minstrell, J., and E.B. Hunt
1990   The Development of a Classroom Based Teaching System Representing Students' Knowledge Structures and Their Processing of Instruction. Phase 1 Report to the James S. McDonnell Foundation.

Minstrell, J., and V. Simpson
1996   A classroom environment for learning: Guiding student's reconstruction of understanding and reasoning. Pp. 175-202 in *Innovations in Learning: New Environments for Education*, L. Schauble and R. Glaser, eds. Mahwah, NJ: Lawrence Erlbaum Associates.

Moats, L.C.
1994   The missing foundation in teacher education: Knowledge of the structure of spoken and written language. *Annals of Dyslexia* 44:81-101.

Moats, L.C., and G.R. Lyon
1996   Wanted: Teachers with knowledge of language. *Topics in Language Disorders* 16(2):73-86.

Moses, R., and C. Cobb
2001   *Radical Equations: Math Literacy and Civil Rights*. Boston: Beacon Press.

Nathan, M.J., A.C. Stephens, D.K. Masarik, M.W. Alibali, and K.R. Koedinger
2002   Representational fluency in middle school: A classroom based study. In *Proceedings of PME-NA XXXIII*, The North American Chapter of the International Group for the Psychology of Mathematics Education.

National Center for Education Statistics
1999   *Highlights from TIMSS*, C. Calsyn, P. Gonzales, and M. Frase, eds. Rep. No. 1999081. Washington, DC: U.S. Department of Education.

2000   *Science Highlights: The Nation's Report Card*. Washington, DC: U.S. Department of Education.

2001   *Fourth Grade Reading Highlights 2000: The Nation's Report Card*. Washington, DC: U.S. Department of Education.

2003   *The Nation's Report Card: Science 2000*, C.Y. O'Sullivan, M.A. Lauko, W.S. Grigg, J. Qian, and J. Zhang, eds. NCES 2003-453. Washington, DC: U.S. Department of Education.

National Council of Teachers of Mathematics
2000   *Principles and Standards for School Mathematics*. Reston, VA: National Council of Teachers of Mathematics.

National Institute of Child Health and Human Development
2000   *Teaching Children to Read: An Evidence-Based Assessment of the Scientific Research Literature on Reading and Its Implications for Reading Instructions*. Report of the National Reading Panel. NIH Publication No. 00-4769. Washington, DC: U.S. Department of Education.

National Research Council
1998   *Preventing Reading Difficulties in Young Children*. Committee on the Prevention of Reading Difficulties in Young Children. C.E. Snow, M. Burns, and M. Griffin, eds. Washington, DC: National Academy Press.

1999   *How People Learn: Bridging Research and Practice*. Committee on Learning Research and Educational Practice. M.S. Donovan, J.D. Bransford, and J.W. Pellegrino, eds. Washington, DC: National Academy Press.

2000   *How People Learn: Brain, Mind, Experience, and School.* Committee on Developments in the Science of Learning. J.D. Bransford, A.L. Brown, and R.R. Cocking, eds. Washington, DC: National Academy Press.

2001a   *Adding It Up: Helping Children Learn Mathematics.* Mathematics Learning Study Committee. J. Kilpatrick, J. Swafford, and B. Findell, eds. Washington, DC: National Academy Press.

2001b   *Eager to Learn: Educating Our Preschoolers.* Committee on Early Childhood Pedagogy, Commission on Behavioral and Social Sciences and Education. B.T. Bowman, M.S. Donovan, and M.S. Burns, eds. Washington, DC: National Academy Press.

2001c   Knowing What Students Know: The Science and Design of Educational Assessment. Committee on the Foundations of Assessment. J.W. Pellegrino, N. Chudowsky, and R. Glaser, eds. Washington, DC: National Academy Press.

2001d   *New Horizons in Health: An Integrative Approach.* Committee on Future Directions for Behavioral and Social Sciences Research. B.H. Singer and C.D. Ryff, eds. Washington, DC: National Academy Press.

2002a   *Minority Students in Special and Gifted Education.* Committee on Minority Representation in Special Education. M.S. Donovan and C.T Cross, eds. Washington, DC: The National Academies Press.

2002b   *Scientific Research in Education.* Committee on Scientific Principles for Education Research. R.J. Shavelson and L. Towne, eds. Washington, DC: The National Academies Press.

2003   *Strategic Education Research Partnership.* Committee on Strategic Education Research Partnership. M.S. Donovan, A.K. Wigdor, and C.E. Snow, editors. Washington, DC: The National Academies Press.

Forth- *How Students Learn: History, Math, and Science in the Classroom.* Com-
coming   mittee on How People Learn, A Targeted Report for Teachers. J.D. Bransford and M.S. Donovan, eds. Washington, DC: The National Academies Press.

Nelson, J., and L. Bader

2001   Physics teaching resource agent program. *The Physics Teacher* 39.

O'Connor, R.E.

1999   Teachers learning ladders to literacy. *Learning Disabilities Research and Practice* 14(4):203-214.

O'Connor, R.E., and J.R. Jenkins

1999   The prediction of reading disabilities in kindergarten and first grade. *Scientific Studies of Reading* 3:159-197.

Olson, D.R.

1977   From utterance to text: The bias of language in speech and writing. *Harvard Educational Review* 47:257-281.

Palincsar, A.S.

1986   The role of dialogue in providing scaffolded instruction. *Educational Psychologist* 21(1 and 2):73-98.

Palincsar, A.S., and A.L. Brown

1984   Reciprocal teaching of comprehension-fostering and comprehension-monitoring activities. *Cognition and Instruction* 1:117-175.

Palincsar, A.S., D.D. Stevens, and J.R. Gavelek

1989   Collaborating with teachers in the interest of student collaboration. *International Journal of Educational Research* 13:41-53.

Pasero, S.

2003   *The State of Physics-First Programs.* A report for Project ARISE (American Renaissance in Science Education). Batavia, IL: Fermilab Education Department.

Pearson, P.D., and D.N. Hamm
2002   *The Assessment of Reading Comprehension: A Review of Practice—Past, Present, and Future.* Santa Monica, CA: RAND Corporation.

Pelavin, S.H., and M. Kane
1990   *Changing the Odds: Factors Increasing Access to College.* New York: College Board Publications.

Perfetti, C.A.
1992   The representation problem in reading acquisition. Pp. 145-174 in *Reading Acquisition,* P.B. Gough, L.C. Ehri, and R. Treiman, eds. Hillsdale, NJ: Lawrence Erlbaum.

Petrosino, A., R. Lehrer, and L. Schauble
2003   Structuring error and experimental variation as distribution in the fourth grade. *Mathematical Thinking and Learning* 5(2 and 3):131-156.

Pfundt, H., and R. Duit
1991   *Student's Alternative Frameworks and Science Education* (3rd edition). Keil, Germany: Institute for Science Education.

Porter, A.C.
1998   The effects of upgrading policies on high school mathematics and science. Pp. 123-172 in *Brookings Papers on Education Policy,* D. Ravitch, ed. Washington, DC: Brookings Institution.

Pressley, M., C.J. Johnson, S. Symons, J.A. McGoldrick, and J.A. Kurita
1989   Strategies that improve children's memory and comprehension of text. *Elementary School Journal* 90(1):3-32.

Raghavan, K., R. Glaser, and M.L. Sartoris
1995   MARS, a computer-supported middle-school science curriculum to foster model-based analytical reasoning. Pp. 17-56 in *Proceedings from the Working Conference on Applications of Technology in the Science Classroom.* Columbus, OH: The National Center for Science Teaching and Learning.

RAND
2002a *Toward an R and D Program in Reading Comprehension.* RAND Reading Study Group, C. Snow, Chair. Santa Monica, CA: RAND.
2002b *Mathematical Proficiency for All Students: Toward a Strategic Research and Development Program in Mathematics Education.* RAND Mathematics Study Panel, D. Loewenberg Ball, Chair. DRU-2773-OERI. Santa Monica, CA: RAND.

Reeve, J., E. Bolt, and Y. Cai
1999   Autonomy-supportive teachers: How they teach and motivate students. *Journal of Educational Psychology* 91(3):537-548.

Reiser, B.J., I. Tabak, W.A. Sandoval, B. Smith, F. Steinmuller, and T.J. Leone
2001   Bguile: Strategic and conceptual scaffolds for scientific inquiry in biology classrooms. In *Cognition and Instruction: Twenty Five Years in Progress,* S.M. Carver and D. Klahr, eds. Mahwah, NJ: Erlbaum.

Roseman, J.E., S. Kesidou, L. Stern, and A. Caldwell
1999   Heavy books light on learning: AAAS project 2061 evaluates middle grades science textbooks. *Science Books and Films* 35(6).

Rosenshine, B., and C. Meister
1994   Reciprocal teaching: A review of the research. *Review of Educational Research* 64(4):479-530.

Rosenshine, B., C. Meister, and S. Chapman
1996   Teaching students to generate questions: A review of the intervention studies. *Review of Educational Research* 66(2):181-122.

Ruiz-Primo, M.A., and R.J. Shavelson

1996   Rhetoric and reality in science performance assessments: An update. *Journal of Research in Science Teaching* 33(10):1045-1063.

Ruiz-Primo, M.A., S.E. Schultz, M. Li, and R.J. Shavelson

2001   Comparison of the reliability and validity of scores from two concept-mapping techniques. *Journal of Research in Science Teaching* 38(2):260-278.

Salinger, T., and E. Chittenden

1994   Analysis of an early literacy portfolio: Consequences for instruction. *Language Arts* 71:446-452.

Scarborough, H.S.

1989   Prediction of reading disability from familial and individual differences. *Journal of Educational Psychology* 81(1):101-108.

1998   Early identification of children at risk for reading disabilities: Phonological awareness and some other promising predictors. Pp. 77-121 in *Specific Reading Disability: A View of the Spectrum*, B.K. Shapiro, P.J. Accardo, and A.J. Capute, eds. Timonium, MD: York Press.

Schmidt, W.H.

2001   Defining teacher quality through content: Professional development implications from TIMSS. Pp. 141-164 in *Professional Development Planning and Design: Issues in Science Education*, J. Rhonton and P. Bowers, eds. Arlington, VA: National Science Teachers Association.

Schofield, J.W., D. Evans-Rhodes, and B.R. Huber

1990   Artificial intelligence in the classroom: The impact of a computer-based tutor on teachers and students. *Social Science Computer Review* 8(1):24-41.

Share, D.L., A.F. Jorm, R. Maclean, and R. Matthews

1984   Sources of individual differences in reading acquisition. *Journal of Educational Psychology* 76(6):1309-1324.

Smith, C.L., D. Maclin, C. Houghton, and M.G. Hennessey

2000   Sixth-grade students' epistemologies of science: The impact of school science experiences on epistemological development. *Cognition and Instruction* 18(3):285-316.

Solano-Flores, G., and R.J. Shavelson

1997   Development of performance assessments in science: Conceptual, practical, and logistical issues. *Educational Measurement: Issues and Practice* 16(3):16-24.

Solomon, D.H., L. Van Houten, R.J. Glynn, L. Baden, K. Curtis, H. Schrager, and J. Avorn

2001   Academic detailing to improve use of broad-spectrum antibiotics at an academic medical center. *Archives of Internal Medicine* 161(August 13/27):1897-1902.

Spillane, J.P.

2000   Cognition and policy implementation: District policymakers and the reform of mathematics education. *Cognition and Instruction* 18(2):141-179.

Stecher, B.M., S.P. Klein, G. Solano-Flores, D. McCaffrey, A. Robyn, R.J. Shavelson, and E. Haertel

2000   The effects of content, format, and inquiry level on performance on science performance assessment scores. *Applied Measurement in Education* 13(2):139-160.

Stigler, J.W., P. Gonzales, T. Kawanaka, S. Knoll, and A. Serrano
  1999   The TIMSS videotape classroom study: Methods and findings from an exploratory research project on eighth-grade mathematics instruction in Germany, Japan, and the United States. *Education Statistics Quarterly* 1(2):109-112.

Stokes, D.E.
  1997   *Pasteur's Quadrant: Basic Science and Technological Innovation.* Washington, DC: Brookings Institute Press.

Strauss, R.P., and E.A. Sawyer
  1986   Some new evidence on teacher and student competencies. *Economics of Education Review* 5(1):41-48.

Thernstrom, A., and S. Thernstrom
  in press *Getting the Answers Right: The Racial Gap in Academic Achievement and How to Close It.* New York: Simon & Schuster.

Thomas, G., S. Wineburg, P. Grossman, O. Myhre, and S. Woolworth
  1998   In the company of colleagues: An interim report on development of a community of teacher learners. *Teaching and Teacher Education* 14(1):21-32.

Thornton, R.K., and D.R. Sokoloff
  1998   Assessing student learning of Newton's laws, the force and motion conceptual evaluation and the evaluation of active learning laboratory and lecture curricula. *American Journal of Physics* 66:338-352.

Torgesen, J.K.
  2002   The prevention of reading difficulties. *Journal of School Psychology* 40(1):7-26.

Turner, J.C.
  1995   The influence of classroom contexts on young children's motivation for literacy. *Reading Research Quarterly* 30(3):410–441.

Valdez-Menchaca, M.C., and G.J. Whitehurst
  1992   Accelerating language development through picture book reading: A systematic extension to Mexican day-care. *Developmental Psychology* 28:1106-1114.

Valverde, G.A., and W.H. Schmidt
  1997   Refocusing U.S. math and science education. *Issues in Science and Technology* 14(2):60-67.

Vellutino, F.R., D.M. Scanlon, and G.R. Lyon
  2000   Differentiating between difficult-to-remediate and readily remediated poor readers: More evidence against the IQ-achievement discrepancy definition of reading disability. *Journal of Learning Disabilities* 33:223-238.

Vosniadou, S., and W.F. Brewer
  1989   The Concept of the Earth's Shape: A Study of Conceptual Change in Childhood. Unpublished paper, Center for the Study of Reading, University of Illinois, Champaign.
  1994   Mental models of the day/night cycle. *Cognitive Science* 18:123-183.

Wagner, R.K., J.K. Torgesen, N.P. Laughon, K. Simmons, and C.A. Rashotte
  1993   Development of young readers' phonological processing abilities. *Journal of Educational Psychology* 85:83-103.

Wagner, R.K., J.K. Torgesen, C.A. Rashotte, S.A. Hecht, T.A. Barker, S.R. Burgess, J. Donahue, and T. Garon
  1997   Changing relations between phonological processing abilities and word-level reading as children develop from beginning to skilled readers: A 5-year longitudinal study. *Developmental Psychology* 33:468-479.

Warren, B., and A. Rosebery
  1996   This question is just too, too easy: Perspectives from the classroom on accountability in science. Pp. 97-125 in the *Contributions of Instructional Innovation of Understanding Learning*, L. Schauble and R. Glaser, eds. Mahwah, NJ: Erlbaum.
Weiss, I.R., E.R. Banilower, K.C. McMahon, and P.S. Smith
  2001   *Report of the 2000 National Survey of Science and Mathematics Education.* Chapel Hill, NC: Horizon Research.
Weiss, I.R., E.R. Banilower, C.M. Overstreet, and E.H. Soar
  2002   *Local Systemic Change Through Teacher Enhancement: Year Seven Cross-Site Report.* Chapel Hill, NC: Horizon Research, Inc.
West, J., K. Denton, and L.M. Reaney
  2001   *The Kindergarten Year: Findings from the Early Childhood Longitudinal Study, Kindergarten Class of 1998-99, Fall 1998.* Washington, DC: National Center for Education Statistics.
White, B.Y.
  1993   ThinkerTools: Causal models, conceptual change, and science education. *Cognition and Instruction* 10(1):1-100.
White, B.Y., and J.R. Frederiksen
  1998   Inquiry, modeling, and metacognition: Making science accessible to all students. *Cognition and Instruction* 16(1):3-118.
  2000   Metacognitive facilitation: An approach to making scientific inquiry accessible to all. Pp. 331-370 in *Inquiring in Inquiry Learning and Teaching in Science*, J. Minstrell and E. van Zee, eds., Washington, DC: American Association for the Advancement of Science.
Whitehead, A.N.
  1929   *The Aims of Education.* New York: Macmillan.
Whitehurst, G.J., and C.J. Lonigan
  2001   Emergent literacy: Development from prereaders to readers. Pp. 11-29 in *Handbook of Early Literacy Research*, S.B. Neuman and D.K. Dickinson, eds. New York: Guilford Press.
Whitehurst, G.J., D.H. Arnold, J.N. Epstein, A.L. Angell, M. Smith, and J.E. Fischel
  1994   A picture book reading intervention in daycare and home for children from low-income families. *Developmental Psychology* 30:679-689.
Wiggins, G., and J. McTighe
  1998   *Understanding by Design.* Alexandria, VA: Association for Supervision and Curriculum Development.
Wiser, M.
  1995   Use of history of science to understand and remedy students' misconceptions about heat and temperature. Pp. 23-38 in *Software Goes to School: Teaching for Understanding with New Technologies*, D.N. Perkins, J.L. Schwartz, M.M. West, and M.S. Stone, eds. New York: Oxford University Press.
Wood, F., D. Hill, and M. Meyer
  2001   *Predictive Assessment of Reading.* Winston-Salem, NC: Wake Forest University School of Medicine.
Wood, T., and P. Sellers
  1997   Deepening the analysis: Longitudinal assessment of a problem-centered mathematics program. *Journal for Research in Mathematics Education* 28:163-186.

A Assessment

While many themes are woven into the proposed SERP research agenda, *assessment* of the outcomes of learning and instruction merits special attention. High-quality evidence that permits practitioners, researchers, and policy makers to ask and answer critical questions about the outcomes of learning and instruction—what students know and are able to do—is critical to advancing a SERP R&D agenda in any domain of instruction.

THREE BROAD PURPOSES

There are three broad purposes for educational assessment:

1. *Formative assessment for use in the classroom to assist learning.* As described in the chapter, teachers need assessment data on their students to guide the instructional process.
2. *Summative assessment for use at the classroom, school, or district level* to determine student attainment levels. These include tests at the end of a unit or a school year to determine what individual students have achieved.
3. *Assessment for program evaluation* used in making comparisons across classrooms or schools. These assessments include standardized tests to determine the outcomes from different instructional programs or to compare performance across schools.

The first requirement for developing quality assessments is

that the concepts and skills that signal progress toward mastery of a subject be understood and specified. In various areas of the curriculum, such as early reading, early mathematics, and high school physics, substantial work has already been done in this regard. In some cases researchers have capitalized on such knowledge to develop the elements of an assessment strategy, although that work has generally concentrated on the development of materials for formative assessment.[1] In contrast, research and theory have not been used to develop similarly valid assessment tools for many other areas of mathematics, for reading comprehension, or for numerous aspects of elementary and middle school science. These design and development activities constitute part of the prospective R&D agendas we have outlined for each separate content domain.

Assessment of the overall outcomes of instruction—summative assessment—is important to the R&D agenda because it allows testing of program effectiveness. But it is important more broadly because the content of those assessments can drive instructional practice. The Force Concept Inventory in physics (see Chapter 4) illustrates the potential for a summative assessment tool based on cognitive and instructional research to have a powerful, positive impact on the redesign of instruction. It has served simultaneously as an evaluation tool to determine the effectiveness of a new instructional approach. In most instructional areas, however, little progress has been made in developing assessment tools that support instruction in this way.

Assessment of the impact of long-term programs of R&D, such as those that would be supported by SERP, is also important. For decision-making purposes, the public policy makers need information to determine the return on investing in an enterprise such as SERP.[2]

---

[1] This work includes assessment of components of early reading (see Chapter 2), development of the Number Knowledge Test and integration into the Number Worlds instructional program (see Chapter 3), and work on conceptual understanding in physics, which is incorporated into the Diagnoser software tool (see Chapter 4).

[2] Although we focus in this report on learning outcomes, for public policy purposes data are also needed on the costs of achieving those outcomes. The point of bringing together work on teaching, learning, organization, and policy in the SERP context is to ensure that knowledge is available in all these domains to support decision making.

The education system generally fails to distinguish the requirements of formative, summative, and program evaluation assessments. What is needed is not only greater sophistication in designing assessments to better serve specific purposes, but also coordination within and between the levels of assessment. A well-designed assessment system would allow for the bidirectional flow of information among the levels.

Current large-scale standardized tests used by most states to assess academic achievement fall short in important respects (National Research Council, 2001c). The models of learning and measurement underlying such tests are generally shallow, raising doubts about the quality of the evidence they can provide about student learning or the impact of instructional programs. If it is to support student learning and provide reliable measures of program effectiveness, SERP must undertake research on, and development of, informative and coordinated assessment systems. The SERP program affords a unique opportunity to pursue research and development on integrated assessment systems because it will involve projects and individuals who are concerned with the range of assessment purposes. Research and development initiatives appear in the agenda for all three subjects. This appendix discusses the common elements of those initiatives.

• •

ELEMENTS OF AN ASSESSMENT R&D AGENDA

Regardless of their purpose, quality assessment instruments depend on the same three components: (1) theories and data about content-based cognition that indicate the knowledge and skills that should be tested, (2) tasks and observations that can provide information on whether the student has mastered the knowledge and skills of interest, and (3) qualitative and quantitative techniques for scoring responses that capture fairly the differences in knowledge and skill among the students being tested. Research and development related to each of the three components is needed in order for assessments to provide reliable indicators of student achievement. For example, researchers have developed sophisticated models of student cognition in various areas of the curriculum, but in many cases this has not been translated into sets of tasks that can be used for assessment purposes. Even in subject domains for which characteristics of expertise have been identified, the understanding of patterns of growth required for assessment purposes, which would enable one to identify landmarks on the way to competence, is often lacking.

To develop assessments that are fair—that are comparably valid across different groups of students—it is crucial that patterns of learning for different populations of students be studied. Much of the development of cognitive theories has been conducted with restricted groups of students (i.e., mostly middle-class whites). In many cases it is not clear whether current theories of developing knowledge and expertise apply equally well with diverse populations of students, including those who have been poorly served in the education system, underrepresented minority students, English-language learners, and students with disabilities. While there are typical learning pathways, often there is not a *single* pathway to competence. Furthermore, students will not necessarily respond in similar ways to assessment probes designed to diagnose knowledge and understanding. These kinds of natural variations among individuals need to be better understood through empirical study and incorporated into the cognitive models of learning that serve as a basis for assessment design.

Sophisticated models of learning must be paired with methods of eliciting responses from students that effectively reveal what they know, as well as tools for comparing and scoring those responses. Current measurement methods offer greater potential for drawing inferences about student competence than is often realized (National Research Council, 2001c). It is possible, for example, to characterize student achievement in terms of multiple aspects of proficiency rather than a single score; chart students' progress over time, instead of simply measuring performance at a particular point in time; deal with multiple paths or alternative methods of valued performance; model, monitor, and improve judgments based on informed evaluations; and model performance not only at the level of students, but also at the levels of groups, classes, schools, and states.

Much remains to be done, however, to improve the use of assessment in practice. Iterative cycles of research and development will be required to capture critical dimensions of knowledge in assessment tools and protocols that can be used effectively by those who have limited psychometric expertise. Research must explore ways that teachers can be assisted in integrating new forms of assessment into their instructional practices and how they can best make use of the results from such assessments. It is particularly important that such work be done in close collaboration with practicing teachers who have varying backgrounds and levels of teaching experience.

This iterative work on new forms of assessment must explore their accessibility to teachers and practicality for classroom use, and their efficiency in large-scale testing contexts. For policy purposes, it is particularly important to study how new forms of assessment affect student learning, teacher practice, and education decision making. Also to be studied are ways that school structures (e.g., length of time of classes, class size, and the opportunity for teachers to work together) impact the feasibility of implementing new types of assessments and their effectiveness. A SERP network of field sites makes the pursuit of such an agenda possible.

Biographical Sketches of Panel Members and Staff

James W. Pellegrino *(Chair)* is liberal arts and sciences distinguished professor of cognitive psychology and distinguished professor of education at the University of Illinois at Chicago. His research and development interests focus on children's and adult's thinking and learning and the implications of cognitive research and theory for assessment and instructional practice. Recently he served as cochair of the National Research Council's Committee on the Foundations of Assessment, which issued the report *Knowing What Students Know: The Science and Design of Educational Assessment*. From 1973 to 1979 he was professor of psychology and a research associate of the University of Pittsburgh's Learning Research and Development Center. From 1979 to 1989 he was professor of education and psychology at the University of California at Santa Barbara, where he also served as chair of the Department of Education. From 1989 to 2001 he was the Frank W. Mayborn professor of cognitive studies at Vanderbilt University, where he also served as codirector of the Learning Technology Center and as dean of Vanderbilt's Peabody College of Education and Human Development. He has a B.A. from Colgate University with a major in psychology and M.A. and Ph.D. degrees in experimental and cognitive psychology from the University of Colorado.

John R. Anderson is professor of psychology and computer science at Carnegie Mellon University. His current research involves the acquisition of cognitive skills and the understanding of how human cognition is adapted to the information processing demands of the environment. He has developed the ACT-R production system and applied it to various domains of memory, problem solving, and visual information processing.

He has published widely on human associative memory, language, memory, cognition, and the adaptive character of thought. He has a Ph.D. from Stanford University.

Deborah Loewenberg Ball is a professor of educational studies at the University of Michigan. Her work as a researcher and teacher educator draws directly and indirectly on her long experience as an elementary classroom teacher. With elementary school mathematics as the main context for the work, Ball studies instruction, professional education, and teacher learning. Her work also examines efforts to improve teaching through policy, reform initiatives, and teacher education. Ball's publications include articles on teacher learning and teacher education, the role of subject-matter knowledge in teaching and learning to teach, endemic challenges of teaching, and the relations of policy and practice in instructional reform. She is a member of the National Research Council's Mathematical Science Education Board and its Division of Behavioral and Social Sciences and Education, and she recently served as a member of the Glenn Commission. Ball has a Ph.D. from Michigan State University.

Jill Harrison Berg is a national board certified teacher on sabbatical from Cambridge Public Schools while serving as a doctoral fellow at Harvard University Graduate School of Education. Her research interest is improving teacher practice and student learning through reflection. Her work in teacher education extends from giving workshops for pre-service teachers, consulting with school teams about implementation of best practices, presenting workshops at local and national conferences, to supporting teacher candidates for national board certification. She is committed to supporting work that recognizes the importance of the practitioner's perspective in developing educational endeavors and has collaborated on special projects with many organizations, including Interstate New Teacher Assessment and Support Consortium (INTASC), Project Zero, WGBH, UNICEF, TERC investigations, and the Massachusetts Department of Education. She is the author of a book on improving the quality of teaching through national board certification.

Susan Carey recently joined Harvard University as a professor of psychology. Previously she was a professor of psychology at

New York University. Her major interests include infant cognition, cognitive development, and conceptual change in childhood. Her research concerns the evolutionary and ontogenetic origins of human knowledge in a variety of domains: number, lexical semantics, physical reasoning, and reasoning about intentional states. Carey is affiliated with the Society for Research in Child Development, the Society for Philosophy and Psychology, and the International Society for Infancy Studies. She has received several fellowships and honors, including the Guggenheim, 1999; the Nicod prize (Paris, 1998); the George A. Miller Lecturer (Society for Cognitive Neuroscience, 1998); a Cattell fellowship; and a Fulbright fellowship. She has a Ph.D. from Harvard University.

Stephen J. Ceci holds a lifetime endowed chair in developmental psychology at Cornell University. He studies the development of intelligence and memory and is the author of approximately 300 articles, books, and chapters. Ceci's past honors and scientific awards include a senior Fulbright-Hayes fellowship and a research career scientist award from the National Institutes of Health. An article he published in *Psychological Bulletin* was awarded the 1994 Robert Chin prize from the Society for the Psychological Study of Social Issues for the best article, and it was named one of the top 20 articles in 1994 by Hertzig & Farber. Ceci has received the IBM supercomputing prize, three senior Mensa Foundation research prizes, and the Arthur Rickter award for his work on children's testimony. He currently serves on seven editorial boards. The American Academy of Forensic Psychology gave Ceci its lifetime distinguished contribution award for 2000, and the American Psychological Association announced he is the recipient of its 2002 lifetime distinguished contribution award for science and society. He recently completed a three-year term on the American Psychological Society's board of directors. Ceci is the coeditor of the journal *Psychological Science in the Public Interest*, which is partnered with *Scientific American*. He is a current member of the advisory committee to the National Science Foundation's Social, Behavioral, and Economics Sciences Directorate and recipient of the American Psychological Association's 2003 award for lifetime contribution to the application of psychology. He is a fellow of the American Association for the Advancement of Science, the Americal Psychological Association, and the American Psychological Society. He has a B.A. from the University of Delaware, an M.A. from the University of Pennsylvania, and a Ph.D. in developmental psychology from the University of Exeter, England.

Mary Ellen Dakin is a secondary school English teacher at Revere High School in Revere, Massachusetts. She has taught in both private and a public school settings since 1987. In 1994, she attended the Teaching Shakespeare Institute at the Folger Shakespeare Library, and since then she has presented workshops on the topic of teaching Shakespeare through performance at regional and national conventions. Her writing has been published in *Shakespeare Magazine* and in the *Harvard Educational Review*. She is a member of the Massachusetts Department of Education's assessment development committee. In 1999, she earned certification in adult/young adult English language arts from the National Board for Professional Teaching Standards and holds the title of master teacher from the Massachusetts Department of Education.

M. Suzanne Donovan (Study Director) is director of the SERP project and a senior program officer at the National Research Council's Division on Behavioral and Social Sciences and Education. She is currently study director for a project that will produce a volume on *How People Learn* targeted to teachers. She was co-editor of the NRC reports *How People Learn: Bridging Research and Practice, Minority Students in Special and Gifted Education,* and *Eager to Learn: Educating Our Preschoolers.* She has a Ph.D. from the University of California at Berkeley School of Public Policy, and was previously on the faculty of Columbia University's School of Public and International Affairs.

Barbara Foorman is professor and director of the Center for Academic and Reading Skills at the University of Texas-Houston Medical School. Her work has focused on reading acquisition and the role of instruction. She is currently principal investigator of a grant from the National Institute of Child Health and Human Development for the study of early interventions for children with reading problems in schools in Houston and Washington, D.C. She has a Ph.D. from the University of California at Berkeley, School of Education.

Walter Kintsch is professor of psychology and director of the Institute of Cognitive Science at the University of Colorado in Boulder. His research focus has been on the study of how people understand language, using both experimental and computational modeling techniques. His current research involves latent semantic analysis, formulating a psychological semantics based on it. Kintsch has received the distinguished scientific contribution award of the American Psychological Association. He has been chair of the governing boards of the Cognitive Science Society and the Psychonomic Society and president of Division 3 of the American Psychological Association. Kintsch has a Ph.D. in psychology from the University of Kansas.

Robert A. Morse has been a teacher of physics at St. Albans School in Washington, DC. since 1982. He is an active member of the American Association of Physics Teachers (AAPT) including being a physics teaching resource agent (PTRA), a workshop presenter to teachers in the District of Columbia, Maryland, and Virginia area private and public schools, and nationally at AAPT meetings and PTRA training sessions. He was a participant in the National Research Council conference How People Learn in 1998. He received the presidential award for excellence in science teaching in 1988. He has a Ph.D. in science education from the University of Maryland at College Park.

Sharon Robinson is president of the Educational Policy Leadership Institute of the Educational Testing Service (ETS) and heads the public policy arm of the company. In this role, she serves as the primary voice for ETS in the national dialogue surrounding education and education reform. Previously she was the ETS executive vice president for external affairs, public policy, and research. Before joining ETS, Robinson was assistant secretary of education in the U.S. Department of Education's Office of Educational Research and Improvement. She also held a variety of leadership positions at the National Education Association, including director of the National Center for Innovation, its research and development arm. A lifelong civil rights activist, Robinson has waged a personal crusade to convince educators of the economic necessity and ethical responsibility to develop strategies for educating and maximizing the potential of minority and disabled students in rural areas and inner-city districts. Robinson has B.A., M.A., and Ph.D. degrees, the latter in educational administration and supervision, from the University of Kentucky. She recently completed the renowned Harvard Business School Advanced Management Program.

Jon Saphier is the founder and president of Research for Better Teaching (RBT), an educational consulting and training organization in Carlisle, Massachusetts, that works with over 100 school districts annually on long-term projects for improving instruction and student learning. A former classroom teacher, staff developer, and administrator, Saphier has developed courses and materials for educators that build on findings from cognitive science and developmental psychology. He is the author of six books and numerous articles dealing with pedagogy, supervision, and school culture. He is also founder and chairman of Teachers 21, a nonprofit organization dedicated to the professionalization of teaching. Teachers 21 focuses on influencing public policy and legislation related to the conditions of teaching and does extensive staff development work with a special focus on beginning teacher induction programs.

Leona Schauble is a professor of education at Vanderbilt University. Her research interests include the relations between everyday reasoning and more formal, culturally supported, and schooled forms of thinking, such as scientific and mathematical reasoning. Her research focuses on such topics as belief change in contexts of scientific experimentation, everyday reasoning, causal inference, and the origins and development of model-based reasoning. Prior to her work at Vanderbilt, she worked at the University of Wisconsin, the Learning Research and Development Center at the University of Pittsburgh, and the Children's Television Workshop in New York. Schauble has a Ph.D. in developmental and educational psychology from Columbia University.

Joseph K. Torgesen, is the Robert M. Gagne professor of psychology and education at Florida State University. His research interests include the psychology of reading and prevention of reading disabilities, cognitive characteristics of children with learning disabilities, assessment practices with children, and computer-assisted instruction in basic academic skills. He is the author or coauthor of over 150 books, book chapters, papers, and tests in these areas. He has been active in numerous professional organizations including as a member of the Learning Disabilities Planning Group, Office of Special Education Programs, U.S. Office of Education; a member of the professional advisory board of the National Center for Learning Disabilities; and a member of the scientific advisory board of the International Dyslexia Association. He has a Ph.D. in developmental and clinical psychology from the University of Michigan.

Alexandra K. Wigdor served as the first director of the National Research Council's SERP project. An NRC staff member since 1978, she most recently held the position of deputy director of the Commission on Behavioral and Social Sciences and Education with special responsibility for developing the education program. Among the notable NRC reports on improving education produced that grew out of that program are *Improving Student Learning: A Strategic Plan for Education Research and Its Utilization* (1999); *Preventing Reading Difficulties in Young Children* (1998); *How People Learn: Mind, Brain, Experience, and School* (1999); *How People Learn: Bridging Research and Practice; Making Money Matter: Financing America's Schools* (1999), and *Eager to Learn: Educating Our Preschoolers* (2000).

Mark R. Wilson is a professor of education at the University of California, Berkeley. His research focuses on educational measurement, item response modeling, and assessment design. His research interests include the incorporation of cognitive modeling perspectives into psychometric models. He has recently concluded a project to develop the BEAR Assessment System, which coordinates assessment information between embedded classroom tasks and more traditional testing methods, in order to better support classroom instruction and educational accountability. He is a founding editor of the journal *Measurement: Interdisciplinary Research and Perspectives*. He is a member of the Joint Committee for Standards in Educational Evaluation. He has a Ph.D. in measurement and educational statistics from the University of Chicago.

Suzanne M. Wilson is a professor of teacher education at the Department of Teacher Education, Michigan State University, and also directs the university's College of Education Center for the Scholarship of Teaching. She is also a senior scholar at the Carnegie Foundation for the Advancement of Teaching. Her areas of expertise include curriculum policy, the history of teachers and teaching, mathematics reform, teacher assessment, teacher education and learning, and teaching history. An educational psychologist with an interest in teacher learning and teacher knowledge, her studies include investigating the capacities and commitments of exemplary secondary school history and mathematics teachers and she has written extensively on the knowledge base of teaching. She is the author of a recent book based on a longitudinal study of the relationship between educational policy and teaching practice in California. Wilson has a Ph.D. from Stanford University.

Index

G

Generalizing
    across studies, 148–149
Goals of school science
    reformulating, 129
Griffin, Sharon, 73

H

Hennessey, Sister Mary Gertrude, 131
Hestenes, David, 112

I

Implementation
    amounts of variability in, 122–123
In-service education, 99
Informal mathematical reasoning
    building on children's, 68
Instructional interventions
    to move students along a learning path, 16
Instructional practices
    promoting reading comprehension, 3–4, 62–63
Instructional programs
    differentiating, 6, 121
Integrated learning-instruction models
    developing and evaluating, 6, 137–138
Integrated reading instruction, 40
    developing and testing reading
        intervention, 46
    learning from exemplary practice, 45
Interdependence
    of student learning, teacher knowledge, and
        organizational environment, 20–24
Investigation
    using methods permitting direct, 146–147
Investigations in Number, Data and Space
    curriculum, 71

K

Kindergarten
    challenge of children entering behind their
        peers, 68
"Knowledge packages," 82
Knowledge-rich goal-focused inquiry
    science as, 130
Knowledge tracing, 94

L

Language development, 32
Learning
    formative assessment in classroom to assist,
        167–168
    trajectory for teachers, 22
"Lift," 133
Linguistic level
    text comprehension involving processing at,
        50

M

Ma, Liping, 77–79, 84
Magic School Bus, 102
Math Trailblazers curriculum, 71
Mathematics, 4–5, 28, 66–101
    algebra, 5, 87–101
    contribution to future earnings, 88
    contribution to test results, 27
    elementary mathematics, 4–5, 66–87
Mazur, Ed, 106
McDermott, Lillian, 104
Meaningful comparisons
    developing assessment instruments to
        anchor, 138–139
Medical metaphor, 11–14
"Mental counting line," 72
Metacognition
    developing, 52–53
Metacognitive strategy instruction, 54, 57
    developing materials for teachers using, 3,
        61–62
Metz, Kathleen, 129–130
Modeling, 94, 131–132
    instruction in high school physics, 112
    introducing physics as, 111
    science as, 133
Models
    of integrated reading instruction, 3, 44–46
Momentum
    misconceptions about, 18

N

NAEP. *See* National Assessment of Educational
    Progress
National Assessment of Educational Progress
    (NAEP), 36, 75, 102
National Center for Education Statistics, 36, 102

National Council of Teachers of Mathematics (NCTM), 82, 95, 99
National Institute of Child Health and Human Development, 10, 31, 38, 44, 46–49, 52, 55, 58
National Reading Panel, 31, 40, 44, 48–49, 54, 58
National Research Council (NRC), 10, 19, 31, 67, 88, 125, 142, 145, 150, 170
    Committee on the Prevention of Reading Difficulties in Young Children, 35
    science standards, 127
National Science Foundation (NSF), 71
    curricula supported by, 86, 97–98, 147
    teacher enhancement projects, 119
NCTM. *See* National Council of Teachers of Mathematics
Newtonian mechanics, 104–105, 118
No Child Left Behind legislation, 43
NRC. *See* National Research Council
NSF. *See* National Science Foundation
Number-knowledge-focused curricula, 43
Number Knowledge Test, 70, 73–75, 168
Number words
    ability to verbally count using, 73
Number Worlds curriculum, 70, 73–75, 86, 168

O

One-on-one adult-child conversation, 43
One-to-one correspondence
    ability to count with, 73
Organizational environment
    hampering adoption and use of improved instructional methods, 22–23
    interdependent with student learning and teacher knowledge, 20–24
Organizing knowledge around core concepts
    subtraction with regrouping, 76–80

P

PALS. *See* Virginia Phonological Awareness and Literacy Screening
Pasteur's quadrant, 11
Payne, Roger, 130
Peabody Individual Achievement Test (PIAT), 75
Phonemic awareness, 34
"Phonics" instruction, 32–33, 38, 40
    analogy phonics, 38
    analytic phonics, 38
    embedded phonics, 38

phonics through spelling, 38
    in student knowledge of early reading, 38
    synthetic phonics, 38
Phonological Awareness and Literacy Screening (PALS), 39
Physics, 5–6, 103–124
    research agenda initiatives, 6, 120–124
    student knowledge, 103–115
    teacher knowledge, 115–120
Physics Education Group, 106–107
Physics teaching resource agent (PTRA) program, 119–120
PIAT. *See* Peabody Individual Achievement Test
Poverty
    and math ability, 75
Practice
    bridging gap with research, 19
    bridging gap with theory, 145
    focus on, 10–14, 21, 29
Pre-service education, 99
Preventing Reading Difficulties in Young Children, 31, 39
Primary school mathematics, 72–75
    ability to count with one-to-one correspondence, 73
    ability to "mentally stimulate" the sensorimotor counting, 73
    ability to recognize quantity as set size, 73
    ability to verbally count using number words, 73
Principles and Standards for School Mathematics. See National Council of Teachers of Mathematics
Procedural fluency, 67
Productive disposition, 67
Professional development programs
    on vocabulary and oral language development, 43
Professional scrutiny and critique
    disclosing research for, 149–150
Proficiency
    mathematical, 67
    needed to meet demands of modern life, 27
Program evaluation
    assessment for, 167–169
Progression of student understanding
    in student knowledge of algebra, 89–91
    in student knowledge of early reading, 31–34
    in student knowledge of physics, 104–110
    in student knowledge of reading comprehension beyond the early years, 51

Teacher understanding
    of assessments, 83
Teaching Children to Read, 31
Teaching number and operations
    evaluating and comparing curricular
        approaches to, 4–5, 85–87
Technology support
    needed to assist teachers, 83
Texas Primary Reading Inventory (TPRI), 39
Text
    assessment of, 55–56
    complexity of, 64–65
Text comprehension involving processing at
        different levels, 50
    linguistic level, 50
    semantic level, 50
    in student knowledge of reading
        comprehension beyond the early years,
        50
    understanding level, 50
Text talk, 53, 61
Theory
    bridging gap with practice, 145
    developing, 145–146
    linking research to relevant, 145–146
Theory building
    science as, 130
ThinkerTools
    reflective assessment in, 113, 116–117, 151
Third International Mathematics and Science
        Study (TIMSS), 102, 128
3-2-1 Contact, 102
TIMSS. *See* Third International Mathematics
        and Science Study
TPRI. *See* Texas Primary Reading Inventory
Transferring strategy use, 54
Travel metaphor, 25–26

U

Understanding
    conceptual, 67
    electrical circuits, 106–107
    fluid/medium effects and gravitational
        effects, 108–109
Understanding level
    text comprehension involving processing at,
        50
University of Arizona, 112
University of Washington
    Physics Education Group, 106–107
U.S. Department of Education, 142

V

Virginia Phonological Awareness and Literacy
    Screening (PALS), 39

W

What children should know and be able to do,
        15
    in knowledge of algebra, 88–89
    in knowledge of early reading, 31
    in knowledge of physics, 103–104
    in knowledge of reading comprehension
        beyond the early years, 49–50
    in knowledge of science across the school
        years, 124–125
    in learning of elementary mathematics, 66–
        67
World-knowledge-focused curricula, 43